# A Passion for Trains

828
828

# A Passion for Trains

## The Railroad Photography of Richard Steinheimer

with text by Jeff Brouws

W. W. Norton & Company
New York   London

For Shirley, who's been my partner in every aspect of my life the past twenty years. —RS

In loving memory of our dear friend Ed Delvers.


Printed in Italy
First Edition

Manufactured by Mondadori Publishing Inc.
Design: For A Small Fee, Inc.

Library of Congress Cataloging-in-Publication Data
Steinheimer, Richard.
A passion for trains : the railroad photography of Richard Steinheimer with an essay/
By Jeff Brouws.
p. cm.
Includes bibliographical references.
ISBN 0-393-057437 (hardcover)
1. Photography of railroads—United States. 2. Railroads—United States—Pictorial works.
3. Steinheimer, Richard. I Brouws, Jeffrey T. II. Title.
TR715.S78 2004
779'.9385'0973—dc22
2004005581

W. W. Norton & Company, Inc.
500 Fifth Avenue, New York, New York 10110
www.wwnorton.com

W. W. Norton & Company, Ltd.
Castle House, 75/76 Wells Street, London, WIT 3QT

1 2 3 4 5 6 7 8 9 0

COVER:
Denver and Rio Grande Western 2-10-2
#1403 with coal drag, Thistle, Utah, 1951

PLATE 1 (FRONTISPIECE):
Colorado & Southern engine house with
SD9 #828, Leadville, Colorado, 1968

# Introduction

PLATE 2
Father and son watching the Southern Pacific's *West Coast* depart, Glendale, California, 1949.

The nucleus around which the artist's intellect builds his work is himself; the central ego, personality or whatever it may be called, and this changes little from birth to death. What he once was, he always is, with slight modification.[1]

—Edward Hopper

THE FIRST TIME I MET DICK STEINHEIMER I was invited to his home, a tract house just off the Bayshore freeway in Palo Alto, California. A road-weary Land Rover back from a trip sat in the driveway, muddied by five states crossed in as many days. It was March 1973 and cold for the Bay Area. I had called several nights earlier, a nervous teenager phoning his hero. Despite my stuttering introduction, "Stein" somehow put me at ease and asked that I visit the following week.

I showed up with my friend Alan Stewart, traveling buddy and fellow disciple. Ushered into Stein's living room, we saw a mountainous array of print boxes strewn about. Precariously balanced, they threatened to spill forth like a photographic Vesuvius. We wandered into an adjacent workroom, where topographic maps showing the Milwaukee Road electrification were push-pinned to one wall like targets for a hunt. Thick red lines traced its route over St. Paul Pass in the Bitterroots of Idaho and Montana. Night photographs of flashed boxcabs in driving rain or Little Joes on skeletal trestles were tacked to the other wall. The room felt like a command post for crafting BIG PLANS. Everything everywhere had something to do with trains. Even then, the intensity of Dick's obsession amazed us.

Back in the front room he poured over 11 x 14s, elucidating details about the Cedar Falls cook shack, the two-dollar-a-night rooms at the railroad hotel in Avery, the intricacies of substation operation and regenerative braking—as if all were familiar to us. And throughout he never stopped talking—his enthusiasm permeated every word. I recall many images of commonplace beauty as he shuffled through the prints: of workers and depots, locomotives and freight trains, the backwater buildings of railroad architecture, photos capturing a sense of place and scene, all surrounded by fragments of enticing western geography. A deep-seated, down-in-the-bones humanity resonated throughout. His passion for what he did, what he saw, was infectious.

And infected we became. Alan and I were to make our own trip into that Milwaukee Road world the following summer. After seeing Steinheimer's images we wanted to leave immediately, so powerful was their imprint upon us.

* * * * *

RICHARD STEINHEIMER'S ARRIVAL INTO the world almost derailed. Born in Chicago in 1929, he and his mother nearly died in a difficult birth. It would be the first of many challenges as he made his way through life. His father, a genius-dreamer, recited Shakespeare, wrote poetry and played numerous musical instruments, but had no skill at staying employed during the Great Depression. After losing what small savings the family had in the crash, and under constant financial strain, Steinheimer's parents divorced in 1935. A poignant leave-taking occurred for the young Steinheimer at Dearborn station when he was six years old. With tickets to Phoenix, and the security of a temporary home with relatives there, Dick, his mother, and sister boarded the *Grand Canyon Limited.* Apparently, his father had come to the station to say good-bye, and rapped on Steinheimer's compartment window in a final farewell gesture as the train departed. It became a pivotal moment, one indelibly etched into the child's consciousness: Dick would only see his father once more during his lifetime. I believe the trauma of this loss is directly responsible for fostering Steinheimer's love of trains; would trains have become so central to him if they hadn't been connected to an absent father? This parental hurt also created a lifelong Steinheimer trait of developing friendships with railroaders he came in contact with. For him, emotional loss, railroading, and a desire for human connection would be forever linked. He simply transferred devotion to his father to the railroaders he met and the trains he photographed.

After what must have been a bittersweet cross-country train trip, the truncated family stayed in Arizona for two months, where Steinheimer's love of the desert bloomed. Relocating to Southern California, around 1939 they settled in a house in Glendale adjacent to the Southern Pacific main line. Every time a whistle approached, Dick was out the door, running trackside to catch a glimpse of cab-forwards or 2-10-2s beyond the backyard fence. Trains ran "through his dreams." Living so close to the tracks became a habit: over the next twenty years Steinheimer lived in five apartments or homes that abutted the busy Southern Pacific main lines of California, their close presence providing solace and continuity throughout his life.

Trains became the bright spot in a childhood otherwise marred by health problems. Weak, sickly, and suffering from asthma and an assortment of allergies, he was a quiet boy who spent most of his time indoors entertaining himself. He lived with his sister, mother, and maternal grandmother, who acted as chief caregiver (and obsessive worrier about his health) but, unfortunately, was emotionally distant from the children. His mother, Frances, while loving and kind, worked long hours and weekends as a secretary for the Army Corp of Engineers. This lack of attention taught Steinheimer to go inward and acquire a heightened degree of self-reliance—a valuable asset integral to his photographic efforts later on.[2]

59
4252

Another source of inspirational strength from his youth was a fireman that lived across the street. The father of Dick's playmate Ken Brooks worked for the Glendale Fire Department and Dick greatly admired him.[3] The elder Mr. Brooks acted as a surrogate father for Dick, and Steinheimer developed an abiding trust in him (and all fireman for that matter, as a child might). This trust, Steinheimer said during a recent interview, was the basis of his decision to obey the command of a senior officer while a photographer in the navy.

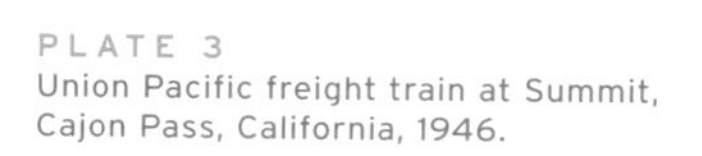

PLATE 3
Union Pacific freight train at Summit, Cajon Pass, California, 1946.

The officer wanted Dick to make aerial photographs of several navy platoons on the parade ground. A hook-and-ladder truck from the base's firehouse was the only way to accomplish this. With the ladder extended one hundred feet, and because Steinheimer trusted firemen (and the sturdiness of their ladders), he bravely scrambled to the top and made the shot, overcoming a deeply ingrained fear of heights. It became an important cornerstone in the foundation of his growing self-confidence.[4] Fear would never again be a part of any photographic activity.

In fact, so many traits Steinheimer exhibited over the course of his five decades of photography—endurance, courage, tenacity, curiosity, intelligence, and a propensity for adventure—bring to mind many of his predecessors. His work was the logical extension of pioneering efforts by western landscape and railroad photographer/explorers like John C. Fremont, who made daguerreotypes in Wyoming's Wind River Canyon in 1842, William Henry Jackson, Andrew Joseph Russell, Timothy O'Sullivan, Alfred A. Hart, Carleton Watkins, Fred Jukes and Lucius Beebe. That he can easily be placed within their ranks, on a continuum extending 160 years, signifies the importance of his lasting contribution to railroad photography in America.

* * * * *

## RUSSELL, JUKES, AND BEEBE'S RESTRAINED EXPRESSION: OR HOW DID ALL THOSE "SMOKING WEDGIES" GET INTO THE STUDIO?

**WHAT ARE HIS ANTECEDENTS?** Who and what came before Steinheimer's groundbreaking work? A discussion of the technical aspects and aesthetics pertaining to railroad photography from 1870 to 1940 is necessary.

Just as Frederick Jackson Turner proclaimed the closing of the frontier in the early 1890s, a concurrent chapter in an earlier era of photography was ending and a new one beginning. Prior to 1888, landscape photographers like Russell or O'Sullivan relied on the collodion wet-plate, which limited their photography to stationary objects and scenes. Exposures were made "by cap" (removing and replacing the lens cap), often lasting from five to fifteen seconds or longer, depending upon the freshness of the collodion and silver baths used for development.[5] Sometimes a primitive drop or flap shutter was used for "instantaneous" exposures (still about two seconds in length), but to truly capture something moving quickly was impossible. Russell's famous photographs of the Golden Spike ceremony at Promontory, Utah, in 1869, and the blurred figures within the frame, indicate his use of long exposures. The turning point occurred with two inventions in 1888: the introduction of dry-plate celluloid film by Eastman Kodak (which was much more light sensitive, thus faster, requiring shorter exposure times); and the focal plane shutter by Ottomar Anschutz of Germany, which enabled photographers to make pictures using shutter speeds of up to 1/1000 of a second.[6]

Action studies of trains in the modern era began earnestly in the 1890s with a bevy of photographers including two Eastern cameramen: F. W. Blauvelt and A. F. Bishop. An early "speed shot" by Blauvelt, taken in 1896, captured the *Pennsylvania Limited* traveling at 53 mph: a phenomenal achievement.[7] Bishop too made action studies around this time near New Haven, Connecticut, using a unique method: his wife standing trackside, outfitted with parasols and white handkerchief, had to signal him to trip the shutter, so precarious was the timing involved in capturing the passing train within the photographic frame.

It is, however, the photography of Fred Jukes—a westerner chronicling the changes taking place in the world of railroading—that is most closely associated with early pictures of trains at speed. Jukes, born presumably in Saskatchewan in 1877, spent his early childhood in Virginia City, Nevada, home of the booming Comstock Lode. Jukes' first photograph—of the Carson and Colorado narrow gauge near Dayton, Nevada, in 1893—was made with a Kodak #2, a simple box camera with a fixed focus 57mm lens. Invented and made available to the American public in 1888 (with slight revisions in 1889), it took 60 or 100 circular exposures.[8] Both film and camera had to be sent to Rochester, New York, for processing. As Kodak's slogan stated: "You take the picture; we do the rest!" One imagines Jukes' response receiving finished 2 1/2″ prints by return mail. It must have been a heady experience. Unfortunately, none of these images remain.

Jukes roamed the intermountain West for the better part of thirty years, working first as a fireman on the Colorado Midland, where his earliest images of the Denver and Rio Grande were made in 1902. Sometime after 1904, when poor eyesight prevented a promotion to engineer, he gave up on the notion of a railroad career, got married, moved to Rawlins, Wyoming, and became part owner in a commercial photography business. He later operated studios in Elko, Nevada, and Blaine and Bellingham, Washington, after 1921, shooting trains all the while.[9] During his Rocky Mountain sojourn, however, he made several visits to Chama, New Mexico, a busy division point on the railroad. Here he committed to film his most treasured narrow gauge photographs: dramatic pictures of smoke-billowing triple-headed 4-6-0s climbing Cumbres Pass between 1907 and 1909. By this time he was employing larger format cameras, most notably a 5 x 7 equipped with a lens capable of shutter speeds up to 1/1800 of a second.[10] Clearly Jukes had no trouble arresting the motion of a passing train. He was so technically proficient and artful one of his well-known action photographs—entitled *A Night's Work Ahead*—became a famous postcard of the day.[11]

While in-depth biographical information about Jukes is scant, it seems his photographic activity ended around 1938. Due to the itinerant nature of his profession, Jukes entrusted the care of his negatives to his brother Henry in Washington. Relegated to an attic, his collection (some speculate this numbered around 250 large plate images) sustained water damage and was subsequently destroyed. Jukes, however, had taken it upon himself to make prints from each exposed negative, later making copy negatives from the originals to make oversized enlargements. Existing prints of his work are few, found only in isolated museum and private collections. Lucius Beebe came into contact with Jukes sometime in the 1940s and, impressed with the composition, clarity, style, and historical significance of his photography, used many of the images in publications he authored over the next twenty years. Beebe, a veritable "vacuum cleaner" for all rail-related photographs, bought the remaining collection in the early 1960s.[12] Jukes' pictures offered insight and direction Beebe's own photography might take. Jukes' striking "at speed" images of Southern Pacific's *Overland Limited,* taken near Elko, Nevada, circa 1916–18 (when he had his studio there)—white smoke receding behind the train like cotton ball clouds—foreshadow later Beebe compositions (see *Age of Steam,* pages 279 and 281, or *The Central Pacific and Southern Pacific Railroads,* pages 208, 211, 214–15, and 226–27).

Jukes, a "lone wolf" rail photographer in the first years of the century, didn't maintain that status for long. In the 1920s, 30s, and 40s, Colorado became a hotbed of activity for camera-toting ferroequinologists (lovers of the iron horse) with many of the country's "pioneer modern"[13] cameramen (Richard Kindig, Otto Perry, and Jackson Thode, among others) shooting there—all perhaps inspired by incomparable scenery and the allure of narrow gauge trains.[14] Steinheimer, too, would find the slim rails and natural beauty of the Columbine state irresistible. He made initial forays into the region in 1953, and soon the narrow gauge lines radiating out of Durango became a sixteen-year obsession—he documented the surreal longevity of this last bastion of western steam through 1969.

* * * * *

IT APPEARS THE ONLY PLACE a young railfan photographer in the 1920s and 30s might see railroad imagery in print was in *Railway and Locomotive Engineering, Locomotive Engineer's Journal,* or *Railroad Man Stories* (which would later become *Railroad)*—a magazine which helped foster the modern railfan movement.

During this time, a small group of collectors were involved in two "schools" of railroad photography. The first group, with Kodak 1A, 2C, and 3A, and later Monitor or Vigilant 616 cameras, made mostly postcard-size images of locomotives and rolling stock. It was all the rage to pick a particular railroad and shoot a picture of each class of locomotive in the favored vantage point of the time: a broadside or 3/4 angle from the left with the side rods down, engine number and road name clearly delineated—preferably in flat light so none of the drive wheel or running gear were lost in shadow (see figure 1). These images ranged from a "snapshot aesthetic"—casual pictures made by amateurs—to a more refined and rigorous approach where all extraneous detail was excluded and the locomotive was centered in the frame. Nothing was to detract from the information the picture conveyed. These types of photos, also called the "locomotive documentary still" or "roster" shot, were routinely made in stations, yards, or roundhouse facilities: places the average railfan had easy access to. The more serious practitioners used sturdy tripods and small lens apertures. Photographs by Gerald Best (who once photographed the entire roster of Southern Pacific locomotives) or the Broadbelt Collection (gleaned from builder photos made by locomotive manufacturers), with their spare, mechanistic beauty, epitomize the best of this genre. The latter, builder prints with masked-out backgrounds, showcase the engine's features, perfectly rendered against a white field (see figure 2). Collectors swapped these two types of prints with fellow enthusiasts, beginning in the 1920s, with some fans amassing 25,000 images or more.[15] After an organization called the International

FIGURE 1
Unidentified "locomotive documentary still" roster portrait of Denver and Rio Grande Western 2-8-2 #460, built by Baldwin in 1903. Photograph taken in 1938 at Salida, Colorado. Author's collection.

FIGURE 2
Unidentified "builder's photo" locomotive roster portrait of a streamlined 4-6-2 with opaqued background. Date and location unknown. Author's collection.

FIGURE 3
Unidentified 3/4 action "smoking wedgie" photograph of a Baltimore and Ohio passenger train. Date and location unknown. Author's collection.

Engine Picture Club was formed in January 1931,[16] the editors at *Railroad* printed the addresses of photographers under each published picture to facilitate this trading. *Trains* followed suit in 1940. Enthusiasts could write one another, and swap or buy prints and negatives for a nominal fee—as little as ten cents or as high as twenty-five cents per item. It was estimated in 1941 that there were "several thousand" people in the United States taking part in this activity.[17]

Swapping was a keen strategy. One assumes hard times during the Depression and gas rationing during WWII made private car travel a luxury, so railfans—without the economic means to journey cross-country to take locomotive pictures of "foreign" roads—could still secure prints to round out their collections. Later, as personal mobility increased and more photographers took to the road to make their own photos, the practice of trading black-and-white prints waned, only to re-emerge with color slides in the 60s. The car, however, would become a key component of the railfan experience in post-WWII America, driving the hobby (and its devotees) literally and aesthetically in new directions (see plate 4).

The second group of photographers were those favoring action photos: pictures of locomotives and trains at speed. Borrowing from the prevailing aesthetic used by "roster" photographers mentioned previously, the same 3/4 angle was employed, because the photographers were still interested in the same information: locomotive and train detail—albeit with smoke, steam, and other indications of movement thrown into the mix (see figure 3). And if the entire train behind the engine was also in the frame, so much the better. The "speed shot" also provided the photographer with a record of his physical engagement with the train—a massive object hurling through space at 65 or 80 or 90 mph. One senses from popular prose of the day that the act of "capturing" a speeding train going by was similar to bagging big game on safari. In the end, however, both still and action photographers were concerned with one thing: documentation, not interpretation. With rare exception few seemed interested in other aspects of the railroad environment. This is an important distinction—one that suggests why Steinheimer broke with established traditions of the day.

The leading proponent of the 3/4 aesthetic in the 1930s, 40s and 50s was Lucius Beebe, whose landmark book *High Iron* (1938) established a new field of publishing: the rail photo book. Florid text coupled with exquisite action studies created a compelling product. *High Iron*, the first of 27 books he authored on the subject, catapulted him (and later his partner, Charles Clegg, who used a Kodak Medalist—the Hasselblad of its day with shutter speeds of up to 1/400 of a second) to modest fame and fortune, with their publishing efforts netting $54,000 in 1952 alone.[18] Many photographers gravitated toward his style perhaps because of the ubiquity of his books in the marketplace. Prior to Beebe, few photographers had attempted speed shots. His adeptness at self-promotion and his social pedigree didn't hurt either. Beebe and Clegg, both members of the East Coast establishment and independently wealthy, often hired chartered trains for book signings, inviting press and railroad dignitaries. He was deeply influential because there were no other influences—for the moment.

Beebe was well educated, a member of New York's Café Society (which he wrote about for the *New York Herald*), and a sophisticate in matters of taste. His lifestyle, while rather rococo, was also steeped in a multigenerational New England conservatism and classicism that may have influenced his pragmatic photographic style. Beebe's style, while it followed established laws about perspective and vanishing points, was also informed obliquely by the latest trends in European modernism, as later images employing rakish angles suggest.

Interestingly, Henry Dreyfuss' redesign of the *20th Century Limited* (a train Beebe was truly enamored with and would later do a book about) occurred the same year Beebe's first book, *High Iron,* was published. The train's promotional posters, created by artist Leslie Regan, depicted not only the train's new streamlined look but showed the whole train in a dramatic three-point vanishing perspective, where the engine loomed in the foreground, practically overrunning the viewer, with the consist receding into an ever-smaller triangular wedge. The posters were similar compositionally to most of Beebe's photographs and may have been another reference point for him photographically, or at least reinforced his aesthetic choices.

It's also been suggested that a close friend of Beebe's and a leading society photographer of the day, Jerome Zerbe (whose work appeared in *The New Yorker* and other publications), was responsible for Beebe's occasional use of radically tilted camera angles for shooting locomotive fronts and architectural details. This was a visual leitmotif coming into vogue on the pages of leading fashion magazines, directly appropriated from Russian constructivism and European modernist photography.

The other major influence on Beebe's photography, albeit somewhat later and more subtly, was Charles Clegg. Clegg studied with Joseph Lootens, a famous photography teacher in New York City, and absorbed lessons calling for straightforward pictures of subjects close to the photographer's heart, with an emphasis on technical prowess and print quality.[19] Under Looten's tutelage, Clegg developed a light, lyrical approach and a good eye for the total railroad scene. An additional influence on Clegg were the bucolic railroad views painted by artist Albert Sheldon Pennoyer (1888–1957).[20] His book *Locomotives in Our Lives* (1954) had a noticeable effect on Clegg's pictorial choices. Like Pennoyer, Clegg pulled back from the train to show more environment, making photos of railroad architecture and trackside paraphernalia, in noticeable contrast to Beebe's elephantine maneuverings focused on motive power. Beebe's later photos—like his shot of a distant smoking Union Pacific train (see page 196, *Age of Steam*) on the Wyoming plains—suggest he was paying attention to Clegg's more moody, atmospheric pictures as well as the work of others.

As no artist or photographer is ever immune to influence, it can be assumed a cross-pollination of styles occurred concurrently within railroad photography during the 1940s and 50s. Beebe was not inured to these artistic forces. Pictures coming across his desk as he prepared manuscripts had to impact his ideas about railroad photography. It would have been easy to recognize the early atypical artistry of a Philip Hastings, Robert Hale, Jim Shaughnessy, or Steinheimer print and be inspired by it.

But for Beebe and other traditionalists—who endeavored to capture the transcendent beauty of steam locomotion accurately—anything short of a 3/4 approach ("the view favored by collectors") was anathema. Beebe's photographic methods were rigorous to a fault, as described in the pages of his book *Highliners:*

> The perfect railroad action photograph—with its rural background, its clarity of definition of all moving parts, its indication of speed through smoke and steam exhaust, its full length view of the train, and its absence of any object or matter to distract the attention from the locomotive and consists themselves—is not easy to come by. There are but a few hours each day when the flat light necessary for the clear depiction of valve motion and wheel arrangement is available, and long distances and inaccessible spots must frequently be achieved to meet these conditions.

This 3/4 speed-shot aesthetic (also known as a "wedgie" or "smoking wedge") may also have evolved from necessity—the camera dictating the pictorial style. The Graflex 3A—produced from 1907 through 1936 with only minor design changes and shutter speeds of 1/10 to 1/1000—was the favored picture-making device for its day. A bulky view camera with a hooded eyepiece above the ground glass, its design forced the photographer to look down at a reversed image. It had to be handheld at waist height or placed on a tripod but it couldn't be raised to eye level—a definite drawback. Imagine the complexity of photographing a moving train at speed, across the murky ground glass, in the opposite direction it was actually traveling. Much film was wasted by photographers who tripped shutters too soon; months of practice were required to wait until it appeared the train was almost too near to take a shot.[21] Standing adjacent to the track, at a 3/4 angle, may have been the easiest way to get a decent image. It wasn't as straightforward as later generations of railroad photographers, utilizing modern lightweight 35mm cameras, derisively assumed. It took finesse. A well-made photograph of a train at speed was a rare accomplishment in 1940.

In addition, the Graflex created other problems. Its vertical focal-plane shutter caused distortions: the top half of a fleeting locomotive sometimes appeared to be leaning ahead of its own pilot. A famous image by Jacques Lartique, made in 1909 of a speeding automobile, humorously took advantage of this photographic faux pas. Cocking the shutter was a two-step process requiring the setting of tension numbers and curtain apertures; the photographer also had to remember to pull the dark slide on a camera back that might not always hold the film plane flat. Clearly, Beebe's pictorial dictates, combined with the Graflex's cumbersome operation, didn't make for unfettered artistic expression. It was photography in a straitjacket.

Then came the 4 x 5 Speed Graphic (popularized by press photographers in the 1940s), and a different game ensued. Now lensmen, through eye-level viewfinders or flip-up wire rectangles that approximated the negative size, could aim at moving trains without difficulty. These advances, and the smaller twin-lens Rolleis or 35mm Leicas—already routinely used at *Life* magazine—made a more free-form image possible, generating aesthetic breakthroughs by photographers worldwide and closer to home. Steinheimer added these necessary tools to his arsenal early on, getting a 35mm Argus in 1946 and a Rolleicord in 1951. While the 2 1/4 x 2 1/4 camera eventually became his mainstay, it wouldn't be until 1963 that his 4 x 5 was relegated to obsolescence. He routinely used whatever camera was handy and had an abiding affection for all formats.

The discipline of the 4 x 5 Speed Graphic also taught Steinheimer valuable lessons, like the importance of getting the shot the first time with economy of film usage. Some of Steinheimer's well-known images (including pictures like plates 7, 60, and 90) were shot with either one sheet or frame of film; and in the case of *Dan Murray, Milwaukee Road conductor, Malden, Washington, 1964* just two (see plate 74). This is in direct contradiction to contemporary documentary photographers who "play" the percentages—bracketing exposures and shooting multiple rolls to ensure success. Time and again, as I perused the 4 x 5 packets containing Steinheimer's life's work at the DeGolyer Library at Southern Methodist University (which my wife and I had occasion to visit for four days in August 2002 and use as a research tool in preparation of this book), I was thunderstruck by how many of his great shots were made in a single exposure. He was fond of the first take and confident in what it contained.

* * * * *

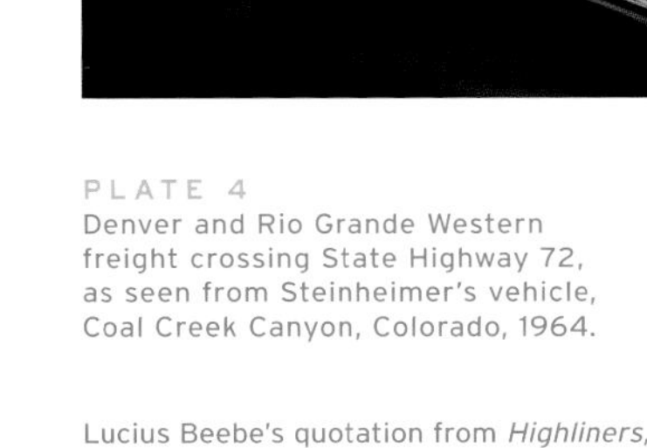

PLATE 4
Denver and Rio Grande Western freight crossing State Highway 72, as seen from Steinheimer's vehicle, Coal Creek Canyon, Colorado, 1964.

Lucius Beebe's quotation from *Highliners*, 1940, courtesy Anne Clegg Holloway.

**PLATE 5**
Southern Pacific's *Starlight* passenger train, Glendale, California, circa 1949.

Lucius Beebe's quotation from *Great Railroad Photographs USA*, 1964, courtesy Anne Clegg Holloway.

THE SECOND IMPORTANT FACTOR shaping railroad photography in America was a man named David P. Morgan. Brilliant with pen and equally adept with visuals, Morgan's sensibilities (with occasional help from in-house art directors and designers at *Trains*) transformed the art form in this country. While his impact was most notably felt after he assumed editorship of the magazine in November 1952, it was also seen earlier while a fledgling writer. A February 1950 *Trains* article entitled "Imaginative insight on the Espee" is a case in point. Morgan selected three unusual photographs of cab-forward locomotives, opening the piece with an off-kilter, time-exposed Steinheimer image heavy on "atmosphere" and light on "information." Morgan waxes:

> This startling night photo of Southern Pacific's *Valley Fast Mail* slamming past Burbank tower ... serves notice that the camera has only begun to record the full sweep and dimension of American railroading. ... What Steinhiemer shot on that October evening in 1948 was not an "effect" or oddity photo but the very real impression that the Espee's hotshot's passage might have left on the mind of any interested bystander. If approached with this thought in mind, Steinheimer's portrait of power ceases to be a blurred and technically poor print and becomes the railroad at night as it is seen and heard and felt. ... In this day of dieselization a photo of such caliber commends itself to that resourceful fraternity of railroad photographers intent upon a new approach to an old subject.

Morgan was an artist, a romantic and realist—influenced by Beebe yet more modern—who wanted to see a broader evocative depiction of railroading. As the up-and-coming arbiter of good taste, Morgan was open to new ideas, as these unpredictable editorial choices show. His vision encompassed the entire railroad scene from shop hand to side-rod to Sherman Hill. Anything railroad-related merited camera time.

But there seemed to be an aesthetic feud, albeit friendly, between proponents of the two differing railroad photography schools (Beebe with his conventional 3/4 wedgies; Morgan with his desire for more expressive interpretations), as suggested by Beebe's sardonic admonishments in the introduction to his book *Great Railroad Photographs USA* (Howell-North, 1964):

> As head-end action shots of trains with smoke exhausts that approached in implication of combustion the Burning of Rome or the Great Fire at Chicago, there began to manifest itself a natural reaction to a type of picture that had become a cliché. The revolt against what he liked to term "miserable wedges of smoke" was spearheaded by David P. Morgan, powerful and authoritative editor of *Trains* magazine, the devotional reading of True Believers everywhere, and at his editorial fiat head-end action suffered a decline and panned action shots taken from parallel-moving motor cars, personnel portraits, trackside atmosphere and train interiors became the preoccupation of photographers who valued his favor.

Beebe's hyperbolic comments concede a revolution had transpired. Under the auspices of Morgan, railroad photography frequently explored fresh territory and achieved new splendors.

Steinheimer was a direct recipient of Morgan's generous, enthusiastic support. Gil Reid, a staff artist at *Trains* (and talented painter in his own right), stated in a recent interview that the magazine's art department deified Steinheimer's photography, so enamored were they with his creativity and prodigious output. From 1948 through 2001, *Trains* published well over 400 of his photographs. They also featured numerous photographic essays; some called "Steinheimer Spectaculars." Examples include: *Santa Margarita Hill* (June 1954), *East Bay* (July 1955), *Arizona Copper Haulers* (July 1956), *An SP California Spectacular* (February 1966), *Action in Utah* (October 1966), *GG1 Curtain Call* (June 1967), *A PA Postlude* (November 1967), *A New Shape in the Snowsheds* (September 1972), *Suspense in a Canyon* (October 1972), *California in the Winter* (February 1974), *Cajon Pass Revisited* (September 1974), *America's Finest Railroad* (October 1974), *Carson & Colorado in California and Nevada* (September 1976), *Tehachapi!* (January 1977), *Mojave Crossing* (August 1977), *The Hill* (December 1981), and *The Joy of Trains* (May 1993). This exposure served photographer and magazine well. Steinheimer's reputation grew to household-name status among the readership, his byline synonymous with leading-edge documentary railroad photography. The "True Believers"/devotees of *Trains* ate up the visuals, as a letter from Morgan to Steinheimer in 1966 attests:

> I never had any thoughts, Dick, but that your one-man SP California "Photo Section" would be popular but even we're surprised at the sheer volume of correspondence which it's attracting. We even had one man who said the balance of the issue was lost on him because he couldn't get his eyes off your prints.

Over the next thirty years, leagues of railfans, captivated by the Steinheimer vision, would elevate his work to iconographic status.

## OF METHODS, MADNESS & "SHUTTER CHANCES": STEINHEIMER'S PHOTOGRAPHY DISCUSSED

So much of every art is an expression of the subconscious, that it seems to me most of all the important qualities are put there unconsciously, and little of importance by the conscious intellect.[22]

—Edward Hopper

IT WOULD APPEAR THAT STEINHEIMER had an innate photographic ability. He had never thought of his photography as art, or considered his efforts serious; in 1955 he called himself "just a rail-photo bum." He was out having fun, indulging a passion. But don't be fooled by this casual demeanor or apparent lack of introspection. He would often go to extreme lengths to get a shot—and liked to "think outside the box" in conceptualizing photographic scenarios. "My best pictures usually emerged from difficult shooting conditions or battles with other adverse factors. I believe that hardship and achievement are part of the same structure—not opposites," he told Freeman Hubbard in a 1978 interview. Steinheimer never minded paying his dues, since his solid effort generally garnered rewarding imagery. He proved this to himself tenfold during his fifty years of capturing trains on celluloid.

Steinheimer's photographs are about the big landscape and testing himself within that landscape. In this regard, Steinheimer's interest are in line with those of A. J. Russell, William Henry Jackson, and Timothy O'Sullivan—all brave, competent, tenacious 19th-century frontier photographers who were thrilled to participate in the dangerous adventure of landmark surveys mapping the West. Steinheimer would have eagerly hired on with the government-sponsored Wheeler or Hayden surveys of the 1870s if he could have, crisscrossing the country on expeditions that explored the Great Basin, Plains, and other new territories of the western United States. The ordeal, the hard-won accomplishments, the dramatically desolate landscapes of those trips would have appealed to him. A word sketch of O'Sullivan by the writer William Kittredge could be a description of Steinheimer as well:

> O'Sullivan proved himself to be tough and determined, capable of first-quality photography, while deep in the rigors of traveling for months in hard country; that he was an expert technician and a developing artist, capable of showing us what he had been seeing in a quite particular way.[23]

And showing us "in a quite particular way" is just what Steinheimer did. After initial attempts with a Baby Brownie behind his house in Glendale in 1945—where he shot the passing parade of Southern Pacific wartime traffic, making classic 3/4 wedgies in the prevailing aesthetic of the day—Steinheimer set his ambitions higher and started making photographs that contextualized the locomotive and train. An awareness of the railroad environment, and the landscape through which trains ran, entered his frame. Beebe's photographic aesthetic, while appreciated by the young Steinheimer (he was, after all, delighted to receive copies of Beebe's two landmark books, *Highball* and *High Iron,* in 1945)[24] was perhaps viewed more as inspiration then something to be imitated. Steinheimer undoubtedly drew encouragement from these books and took comfort in the knowledge that railroad photography was a serious endeavor practiced by other like-minded individuals. But it seems to have occurred to him that Beebe offered only a limited vision of what railroad photography could be, and the field, therefore, was wide open to a more interpretive, creative approach. Beginning in 1946, Steinheimer's prodigious output shows an evolving diversity of imagery and experimentation, covering all aspects of the railroad milieu. As he said recently: "I always wanted to take my photography a little further."

* * * * *

THE AUTOMOBILE, AND THE AUTONOMY it provided, had a tremendous impact on Steinheimer's emergent vision (see plates 4, 57, 61, and 123). Previously he'd only been able to walk to the Glendale Southern Pacific station in daylight from his home about two miles away, with the previously mentioned Kodak Baby Brownie Special or later a 35mm Argus A-2, a camera he first acquired in 1946. But with the loan of his mother's 1939 Chrysler coupe, he was able to pack the more cumbersome 3 1/4 x 4 1/4 Speed Graphic (purchased in 1947) and tripod in the trunk and venture out alone after dark, a time he felt held more promise for dramatic image making than the drab daylight hours. At night, large yard lights danced over the steaming locomotives, producing chiaroscuros of pooled light, obscuring some details, highlighting others, creating high drama in black and white. Occasionally a flashbulb in a reflector was also used to add fill or side lighting. Train after train entered and exited this stage like transitory actors: the *Lark,* the *Coast Daylight,* the *Tehachapi,* all provided excellent performance material for the young photographer's lens. As he discovered the enthralling pleasures of night photography (almost seven years prior to O. Winston Link's herculean nighttime efforts on the Norfolk and Western), Steinheimer's technique and eye were honed. Here he committed to film his earliest signature images.

But Steinheimer wasn't content to stay near home. The flatland geography of the San Fernando Valley—an expansive patch of suburban sprawl (fueled by the emerging post-WWII boom economy) motivated the young Steinheimer to get in his car (or ride the *San Joaquin Daylight* or *San Bernardino Local*) and satisfy his curiosity

FIGURE 4
CB&Q publicity photograph taken in Sheep Canyon, near Greybull, Wyoming, 1940s, as seen in Lucius Beebe's *Highball* by Steinheimer as a young child, courtesy BNSF Railway.

PLATE 6
CB&Q northbound freight with SD9 #448 in Sheep Canyon, near Greybull, Wyoming, 1964.

PLATE 7
Denver and Rio Grande Western 2-10-2 #1403 with coal drag, Thistle, Utah, 1951.

about what lay beyond his youthful purview.

With Cajon and Tehachapi only a few hours away, Steinheimer soon fell in love with the western landscape. He must have decided after initial out-of-state car trips and a train trip to Tucson in 1946 that the majestic lands of the intermountain West would become an integral backdrop for his photography. In fact, one assumes he drew direct inspiration from a specific image in the aforementioned *Highball*—an image of Wyoming's Sheep Canyon by a Chicago, Burlington and Quincy (CB&Q) photographer in the early 1940s (see *Highball*, page 56; figure 4 here). Nineteen years later in 1964 he would duplicate this view (see plates 6 and 89). Ironically, this was one of the few images in *Highball* that showed the train in a bold geology—dominated by canyon and river, flanked by mountainous terrain—not the traditional 3/4, rail-level view normally favored by Beebe. That it might have been an early unconscious touchstone for Steinheimer, an inducement to new directions, is a distinct possibility.

During WW II Steinheimer became very aware of the pictorial magazines and large-format picture books entering American popular culture. He was especially moved by *Life*'s graphic imagery—often splashed across double-page spreads—depicting the tragedy of war, or later, the hardscrabble humanity of a rural physician in Colorado, the cultural depictions of life under Franco, and the fabled career of a famous twentieth-century artist. These images, made by W. Eugene Smith and David Douglas Duncan, two of his early heroes, opened a visual path for the young photographer. Steinheimer was impressed by their ability to create essays where the photographer, and the viewer, became active participants in the story and the way in which one photograph could inform another in the course of telling a story. He cites Duncan's book *The Private World of Pablo Picasso* as an excellent example of photo integration—where all the pictures help carry and advance the narrative line.[25] He was deeply moved by the heart and dogged persistence that resonated throughout Smith's and Duncan's work; the flexible, varied vision they brought to the image-making process; and the personal, resolute commitment to their subjects that was evident in each assignment. These pictures inspired him to gain access behind the scenes, and to delve deeper than most into his subject matter.

Another hero of Steinheimer's was Joe Rosenthal, his teacher at San Francisco City College in 1949. Rosenthal, a combat photographer during WWII, made perhaps the most iconographic image of the conflict: six soldiers raising the flag above Iwo Jima. Steinheimer would later emulate Rosenthal's fortitude in his own photographic pursuits—solitary treks into remote locations in either blistering heat or freezing cold, or dangerous rides atop moving locomotives. If Smith's and Duncan's photography provided an emotional and aesthetic approach for making the pictures, Rosenthal's example went beyond the visual and suggested a certain spirit and mental capacity required to get the job done. All three men's philosophies were like the legs of a tripod in which Steinheimer placed his trust and ambition as a fledgling photographer and as a photojournalist for the *Marin-Independent Journal*, a job he held from 1956 through 1962.

The automobile also provided Dick (and a bevy of friends) with their own *On the Road* experiences, with one friend in particular, Don Sims, playing Sal Paradise to Steinheimer's lanky Dean Moriarty. The train photographers' adventures were no less impressive than those of their literary counterparts. Both buddy teams were enamored with the West, both went to extremes to "make things happen," both courted the edge with insouciance and daring. One oft-discussed trip Steinheimer and Sims made in December 1951 has taken on mythological proportions within the railfan community and bears retelling.

Hunkered down in a dilapidated 1937 Ford lifted from the pages of the *Grapes of Wrath*, Sims and Steinheimer left Los Angeles bound for Utah. Rumors persisted of lingering steam activity on the Denver and Rio Grande Western (D&RGW) near Soldier Summit (the reason for the trip). It was a typical, balmy Southern California day with the temperature hovering at 70 degrees. Twelve hours later, around 3 AM, they stopped at a roadside beanery in Beaver, Utah—weary, hungry, and almost out of gas, the mercury now at 15 below. Ill-prepared for such inclement weather, and in a vehicle with no heat to speak of and no warm clothes, the two heroes struggled against frostbite and a never-ending night. Finding an open service station at dawn, they filled up and pressed on through frigid conditions to arrive at Thistle, Utah, a few hours later in clear, spectacular sunlight. As they approached the helper terminal, plumes of multicolored coal smoke punctuated the sky, announcing the presence of Mallets and 2-10-2s. Checking in with train crews, the two photographers discovered several freights were leaving soon. Sims, bitterly cold, bone-tired and growing impatient, suggested they warm up over coffee in a roadside café before heading out. Steinheimer, sensing something wonderful was about to occur, refused any delay—not an unusual dilemma for two people traveling under adverse conditions.[26] Agreeing to disagree, the two parted company momentarily—Steinheimer drove east of town and parked the car near an open landscape, and Sims trudged off to warm up. Now, many photographers often tell stories about how a picture "got away," or almost did. The image Steinheimer exposed just a few moments later: *28 Degrees Below at Thistle, Utah, 1951*—his first masterpiece—might not have been committed to film if he had agreed to go thaw out instead of continuing on with the photography (see plate 7). For Steinheimer and Sims, despite the travail, the ensuing five days were nirvana—steam locomotives in splendid scenery, gallant in the final throes of their last days. It got no better than this. Overcoming

PLATE 8
Southern Pacific's *Starlight* passenger train, Glendale, California, circa 1951.

harsh conditions triumphantly—with first-rate negatives as evidence of their accomplishment—meant a great deal to both of them.

* * * * *

STEINHEIMER EMPLOYED SEVERAL METHODOLOGIES in his work; some were intentional and conscious, others less so. Most important was his practice of "layering"—covering a specific railroad, locale, or locomotive over the long term, frequently returning to scenes or situations to broaden his coverage on particular themes. Undoubtedly his newspaper background proved influential in this regard. He knew the deeper you went in, the longer you stayed, the better the pictures were—and subsequently the more complete the story became. As he once told a fellow photographer: "We're not just making photos; we're creating a body of work." Repeated trips began enduring love affairs with the Milwaukee Road, the Rio Grande narrow gauge, Southern Pacific's Donner Pass, or the Alco PAs—to name just a few of his passions. Layering also took pressure off the photographer to get every picture needed the first time around—as a self-directed project without specific deadlines will do. It was also a pleasant respite from a stressful but rewarding career as a Silicon Valley industrial photographer—a job he held through most of the 1960s.

Layering could also be practiced on a personal level. A photographer hanging around an engine house or yard eventually becomes a known entity: familiarity breeds openness. Steinheimer spent nearly as much time getting to know railroaders as taking pictures of them. The trains, the people, the railroad environment—all got democratic treatment on his contact sheets. Dick speaks with equal feeling about the death of a roundhouse worker he befriended and photographed, as he does about the scrap yard scenes he witnessed of locomotives under the torch (see plate 43). This empathy gave his photography an emotional quality, and his unusually strong commitment to his material was immediately evident. Steinheimer also made trainmen active elements in his compositions, helping to pioneer (along with Hastings) the inclusion of human interest in rail photography. His pictures of the CB&Q sectionmen (in Wyoming's Wind River Canyon, see *Done Honest & True*, page 54) or the Milwaukee Road conductor at Malden, Washington (see plate 74), are expressive, compassionate documents, on par with some of the best portraiture by Dorothea Lange or Walker Evans.

The railroaders also appreciated a photographer who said he'd send prints and did. This type of follow-through created long-lasting bonds with the workers Steinheimer encountered, and these friendships often gave him behind-the-scenes access. These railroaders (especially on the Milwaukee Road) became the father figures or brothers Dick had longed for; they became people he admired and trusted—people, as he once said "confident enough to be mere people." This camaraderie was as important as the photography, maybe more so.

* * * * *

THE FACT THAT STEINHEIMER DIDN'T STOP SHOOTING railroad subject matter after steam's demise is an important characteristic setting his work apart. Not that he didn't love steam and mourn its passing. He was, as he once wholeheartedly admitted, a railfan "scarred by the 50s." But his enthusiasm for trains never waned because he saw the bigger picture: the world of railroading encompassed more then just motive power. A vast river of source material, its visual tributaries were the workers, architecture, and landscapes the railroad ran through; the weather that impacted train movement; the passenger and freight trains, the short lines, branch lines, and main lines; not to mention the railroad's fascinating history. There was a surfeit of material beyond anyone's ability to exhaust.

Also, unlike many photographers from his era, Steinheimer enthusiastically embraced the diesel locomotive and its aesthetics. This unusual attitude did not go unnoticed. A letter from *Trains and Travels*' Wallace Abbey in 1952 asked Steinheimer to participate in a symposium about how to photograph diesels to be published in an upcoming issue of the magazine:

> how can we [other photographers] put into a picture of a diesel the active appeal that is inherent in a picture of a steam locomotive? We'd like your answer for this problem. We'd like to know what technique you use to make a diesel look alive, to give it the qualities of motion and power that, compared to a steam locomotive, it lacks because its moving parts are covered and because its general appearance is not greatly impressive. We'd like you to advise our other contributing photographers how they can adapt your technique to a general style. And we'd like you to include what you think is your best picture of a diesel from this angle, so that we can illustrate your idea. Of course, because *Trains and Travel* uses no color photography, color can't enter the discussion.

Abbey's letter proves that the introduction of diesels on the railroad scene had undeniable consequences on photographer's pictorial choices—negative and positive—and was another reason railroad photography shifted direction in the 1950s. Photographers had to adapt or stop—which was never an option for Steinheimer; he just kept going. This, again, was a unique philosophical stance for a photographer who'd seen the death of steam and yet chose to go forward and wade into uncharted creative waters.

Whether Steinheimer ever responded to the above letter we'll never know; Abbey insists the article never ran. But if Steinheimer had replied, he might have elaborated on several techniques that

PLATE 9
Union Pacific 4-6-6-4 Challenger pilots a freight at sunset, near Rawlins, Wyoming, 1952.

were evolving under his skillful eye, as creativity for him—even during this formative phase—seemed an inexhaustible commodity. Because diesels didn't belch smoke, didn't have exposed running gear or drive wheels, and, as Abbey states, seemed "generally unimpressive," Steinheimer was drawn to explore the innate properties of photography itself: namely the use of graphic form, content, and scale; the interplay of tonal relationships in a black-and-white print, unusual compositions and camera angles (he was one of the first, in 1956, to make aerial photographs of trains; see page 414 of Lucius Beebe's *Central Pacific and Southern Pacific Railroads*); the optical effects of panning or slow shutter speeds to indicate motion; the impact of weather and darkness on his favorite subject matter; and even the narrative possibilities of the medium. Not that some of these techniques hadn't been used in photographing steam locomotives—they had, but now it seemed the photographer's intellect, emotional insight, and camera work would have to carry the burden of creativity instead of relying solely upon the inherent photogenic qualities of steam locomotion. It was a challenge Steinheimer welcomed as he experimented with myriad photographic effects and approaches over the next twenty or so years.

* * * * *

The contemplation of things as they are, without error or confusion, without substitution or imposture, is in itself a nobler thing then a whole harvest of invention.

—quote tacked on Dorothea Lange's darkroom door, attributed to the painter Francis Bacon

STEINHEIMER WAS ONE OF THE FIRST to move rail photography into an interpretive mode. He wanted to capture the essence of things and place the railroad into a larger physical, emotional, and sociological context; these were intuitive decisions, not necessarily conscious ones. Two dramatic images, as seen in a September 1972 issue of *Trains* on pages 28–29 ("The Diesel Drama") are illustrative. Steinheimer, positioned in deep shadow with a telephoto lens aimed

THE DIESEL DRAMA

HIGH IN THE SIERRA on a June day in 1966, the diesel drama unfolded before a frightened doe, much as it enveloped each of us at some other place and time. For the doe and her mate, Espee Extra 8460 East was a throbbing, enlarging, oncoming noise, an alarm of danger and change. The wildlife scampered toward its hillside thicket, lingering not as the lead SD40, half in silhouette, half bathed in sunlight, roared around the curve near Crystal Lake. The V-16's shouted away in Run 8, and multiples of powered wheels grasped for traction on a grade of 1.9 per cent at an elevation of 5700 feet.

At some other place and in some other time we did not run when Electro-Motive first approached, but surely we shared the awe and bewilderment of this doe. For an older generation, the first EMD we observed brought the beginning of the awful realization that the steam season was not permanent. It is the ultimate compliment to EMD and all its works that another generation equates the diesel with all locomotives; the diesel that generation first witnessed replaced the diesel that replaced the steam engine of older recall.

So on up comes Extra 8460 East, climbing, climbing, climbing to the crest of The Big Hill, grinding out the revenue ton-miles; coming in the wake of AC's; laying down the sand for a more powerful series of SD's yet unborn, spelling out for a frightened doe and her mate the immense consequence of that day in 1922 when H. L. Hamilton and his friends rented office space at 17th and Euclid in Cleveland, Ohio.—D.P.M.

FIGURE 5
Spread from *Trains* magazine, "The Diesel Drama," featuring Steinheimer images, September 1972, courtesy *Trains* magazine.

at a side-lit Southern Pacific main line, waits for an oncoming freight train. The diesels, still a ways off, shatter the silence of Donner Pass and surprise a young doe. The following image shows the freight train's lead locomotive, "half in silhouette, half bathed in sunlight" as written by David P. Morgan, threatening to run us over, its headlamps two lone highlights breaking up the black expanse of its nose. The two photos paired with Morgan's words admirably convey the initial shock experienced by railfans who witnessed diesels supplanting steam, while simultaneously suggesting an inevitable acceptance in weathering this sea change. It's a more complex reading then afforded most train photographs, but Morgan's sensitivity and intelligence—combined with Stein's graphic punch line—wedded the narrative and became a hallmark of many *Trains* frontispieces by this dynamic duo.

Although he claims no interest in "symbolic" photography, many of Steinheimer's images impart deeper meaning. In a hastily written "notes to myself" from 1983, Steinheimer states: "I'm not a great, or maybe even good, symbolic photog. I don't see symbolism or really care. I see the people or situation or reality—that's it"—a typical self-deprecating assessment. He goes on to talk about a photograph by one of his heroes, W. Eugene Smith, entitled *The Walk to Paradise Garden,* and how Smith's other work from Minamata in the 1970s (which chronicled the mercury poisoning of a small fishing village by a large corporation in Japan) or the even earlier Pittsburgh series from the 1950s both towered over the symbolic mawkishness of *Paradise Garden.* Apparently freighted images rankled Steinheimer; he'd brook no pretension in his photography or anyone else's. While Steinheimer may not have been consciously striving for metaphor, no photograph is ever completely devoid of emotion or meaning. He knew the power of the medium and knew that a photograph, or series of photographs—expressing an unvarnished reality—could convey important social themes and messages. *Trains* editor David P. Morgan certainly thought so and routinely put Steinheimer's imagery to work symbolizing major changes in railroading, as "The Diesel Drama" indicates.

* * * * *

HISTORY, AND PERHAPS THE IMAGES of earlier nineteenth-century photographers, also informed Steinheimer's work. Scanning his library you find books on California's past and weighty tomes on western railroads. George Stewart's *Ordeal by Hunger* (the story of the ill-fated Donner Party) and the rich storyline of the Central Pacific construction may have fostered Dick's initial interest in Donner Pass and the Sierra Nevada. Places like Stanford Curve (in Cold Stream Canyon) or Norden's intriguing snowsheds became familiar haunts. He hiked, snow-shoed and camped all over "The Hill" from Roseville to Truckee—perhaps knowing this landscape better than any other living photographer. Donner became for Steinheimer what Yosemite was to Ansel Adams: a place inextricably linked with his name and reputation. That he would have intimately known its history is assured. His friend Chuck Fox confirms that the level of esoteric information Steinheimer possessed about all things rail-related was encyclopedic. Steinheimer had a good working knowledge of most railroad situations he found himself in and often researched places before traveling to them (as his numerous *Trains* articles verify). Looking at his 1964 Milwaukee Road photograph of cowboys herding cattle (see plate 29) taken west of Three Forks, Montana, on the Lewis and Clark Trail, Steinheimer, his voice welling with emotion, recounted the tale of Sacajawea and her pivotal role in aiding the intrepid explorers. He had a deep sense of the historical underpinnings behind his images and knew that individual landscapes had a history prior to his arrival. Pictures made by Steinheimer near Devil's Slide, a landmark originally photographed by Watkins in 1878 and Jackson in 1880 (see figure 6 and plate 10), while riding the UP's Park City branch in Utah in 1953, or his photograph near Castle Rock Butte, Green River, Wyoming (see *Steam's Finest Hour,* page 92, Kalmbach 1959), also attest to his knowledge of earlier geologic discoveries and landforms included in his predecessors' images. The undertaking of his Centennial Project in 1968 (as mentioned below) indicates an awareness of Russell's and Jackson's work as primary visual documents of the transcontinental railroad construction. That Steinheimer felt a kinship with the landscape, the historical record, and the image makers of yesteryear is evident. Like Russell, Jackson, and O'Sullivan, Steinheimer utilized western topographies in his art to express something else about railroads, nature, and their relationship to each another.

FIGURE 6
*The Devil's Slide, Weber Can(y)on, Utah, 1878,* photograph by Carleton Watkins, courtesy Stephen White II Collection, Los Angeles.

PLATE 10
Union Pacific's *Park City Local* at Devil's Slide, Weber Canyon, Utah, 1953.

To this end, he was also one of the first contemporary rail photographers to use the notion of scale—in particular, images of small trains dominated by imposing or barren geography (see plates 24, 25, 32, 71, and 95; see pages 30–31, *Trains,* March 1972). Steinheimer's veneration of the natural world took on similarities to Japanese landscape painting and rail photography (both of which he admired) and their depiction of man's insignificance dwarfed by the benevolent forces of nature. His nineteenth-century predecessors (Jackson, Russell, O'Sullivan, and Watkins) had also used this technique of scale to express their reverence for nature, and its superiority over mankind's works. Russell, who started his career as a panorama painter, based many of his early Union Pacific photographs, found in his portfolio *The Great West Illustrated* from 1869, on seventeenth-century Dutch landscape paintings, employing the "picturesque mode," where "the

PLATE 11
Southern Pacific E9 and *Coast Daylight* shot from Steinheimer's rooftop, Palo Alto, California, 1964.

human element is usually subordinated to the features of the landscape."[27] Steinheimer could have imitated these art-historical motifs.

Steinheimer was also one of the first rail photographers to fully explore weather and darkness as visual elements. He listened intently to weather reports and would go shooting when the climatological conditions promised to be at their worst, photographing in rain, snow, lightning, or fog, because "this is where and when the good pictures happened." A few isolated railroad images were made previously under adverse conditions—like C. R. Lively's memorable snowbound pictures of the D&RG narrow gauge (which have been falsely attributed to Jukes). These, however, were anomalies and not a part of any photographic strategy repeatedly employed by photographers. Steinheimer, singularly it seems, had a penchant to be out in the world taking pictures at odd times when (and where) most people were not—the opposite of most rail photographers, who generally relied on fair weather and never wandered more then a few hundred yards from their vehicles. His wife, Shirley Burman, relates one story about trekking for three hours to a remote location to make photos and, when she asked Dick what the point was of going to such lengths, he responded, "... look behind you, do you see any other photographers here?" Clearly, if Steinheimer had to hike thirteen miles into a site, travel over inhospitable terrain in a mud-encrusted Land Rover, or stay up all night to shoot due to work or familial obligations, that's what happened. To measure the importance of this development is to realize its significant departure, attitude-wise, from the normal railfan habits of the day. Prior to the late 1940s or early 50s only a handful of other photographers working in the United States (Philip Hastings, Jim Shaughnessy, and O. Winston Link among others) ventured out after dark or took photos in inclement weather. Beebe and Clegg, with isolated exception, photographed in favorable conditions, making no shots after sunset. Theirs was a gentlemanly daylight pursuit, Steinheimer's a more nitty-gritty, in your face, 24-7 endeavor. He liked operating outside of predetermined comfort zones and savored the possibilities of bringing back something original: stormy weather or after-midnight sorties often produced stunning results. These images became his visual calling card.

The automobile also enabled Steinheimer to try different techniques, like the pacing shot. Its cousin, the pan shot (literally standing in one spot and "panning" the camera as the train went by) had first been utilized by a photographer in 1911 along the New York Central.[28] With America on the move after WWII, railfans took to the roads in droves, and pacing shots became a natural by-product of this new mobility. This type of picture involved a train, a track, and an adjacent two-lane road running parallel to the railroad. With the car matching the speed of the train, the photographer—employing a semislow shutter speed around 1/60 of a second—could either use the car window as a framing device or shoot beyond the window toward the train. The slow shutter speed blurred the background but was fast enough to keep the locomotives sharp. The successful pacing shot conveyed the notion of speed, power, solidity, and motion—important ingredients in a Steinheimer photograph (see plate 61). While Dick can make no precise claim for its invention, it can be assumed he saw photographs using this technique first by Hale and later by Hastings (in the pages of *Trains*) and added this pacing technique to his growing photographic arsenal. Steinheimer recounts that "Chief" Hale and he met while both were stationed in the U.S. Navy at San Diego in the spring of 1952, literally colliding in the base's darkroom one day. According to Dick's records they made several weekend outings together. Hale, it will be remembered, spent a decade refining this pacing/panning technique, most notably along the Union Pacific in Wyoming (see page 265, *Age of Steam*), and I think it can be presumed that a viewing of each other's prints and a sharing of techniques and information occurred.

* * * * *

IT'S NEVER BEEN HARD TO IDENTIFY a Steinheimer: more often than not an improvised, seat-of-the-pants virtuosity is on display. And like a jazz musician, Steinheimer stepped repeatedly onto the photographic bandstand and "blew." The ability to improvise was a distinguishing attribute of his style: in each new Steinheimer pictorial gracing the pages of *Trains* during the 1950s, 60s, and 70s you might see something you had never seen before, so masterful was he at the unexpected. For him the "now" was a kaleidoscope of opportunity. He sought out unpredictable situations—and if he failed the first time out, he was back two weeks later to try again.

One of these occasional failures led Steinheimer, in September 1964, on another photographic lark that paid off handsomely. From the rooftop of his home in Palo Alto—once again the backyard fence wasn't one hundred feet from the Southern Pacific's Peninsula main line—he repeatedly made handheld pan shots of the late afternoon commuter parade. The first attempts (see plates 11 and 153) seem rough at best, but he pushed himself repeatedly, mounting the camera eventually on a tripod, and finally nailed what he was after—a precisely rendered, sharply focused speeding locomotive against a blurred background—as the pictures taken in July 1965 confirm. The DeGolyer archive reveals that Steinheimer also toyed with shots of commuter trains going through grade crossings, gates down, a phalanx of autos huddled-in-waiting. While several hundred frames were exposed, he never quite seemed to get what he was after. A rarity. But the ultimate expression of his improvisational flair occurred deep in the Bitterroots of Idaho on the Milwaukee Road

SOUTHERN PACIFIC

in 1974, as he made after-dark time exposures under a full moon from the platform of a SD40 aboard a moving freight train looking down onto the nose of a Little Joe electric locomotive (see plate 118). To look at contact sheets of the twenty-four frames he made—to extend the jazz analogy again—would be a visual equivalent of a free-form John Coltrane solo; they fall on the eyes anachronistically, as evidence of the "new." That these images established a new benchmark in railroad photography is undisputable. After almost thirty years of photographic activity, Steinheimer was still operating at the pinnacle of creativity; he still had the power to "wow" the viewer and himself.

* * * * *

PLATES 12–15
Station and siding sign typology: Riddle, Oregon; Smoke Creek, Nevada; Port Chicago, California; Lone Pine, California, 1960s.

PHOTOGRAPHERS WHO DOCUMENT THINGS or places that may soon disappear are well acquainted with the concept of loss. Unlike his nineteenth-century counterparts, who took pictures celebrating the build out of a transcontinental network, the unification of a nation, or glorious promises of industrial growth, Steinheimer instead saw the sheen come off railroading's Golden Age. He witnessed a transitional period: the lapse of passenger service, the demise of steam, waning traffic levels, and the changeover from first- to second generation diesels. Later decades saw even further attrition within the industry, as methods of operation disappeared: the caboose, the interlocking tower, and the telegrapher. Luckily, Steinheimer found himself uniquely positioned to chronicle these changes, these *petit mortes*, which contributed to the narrative possibilities in his photographs. They also gave him ample opportunity to document the evanescent nature of America's postwar railroad culture.

Steinheimer's copious notes from 1964 through 1966 suggest a keen awareness of the changes occurring in railroading and his desire to "seize the day" before the opportunity passed. Subconsciously he knew what every photographer eventually realizes: that his pictures would one day be portals onto a world that no longer existed. To this end Steinheimer indulged the quirky habit of amassing collections of both station signs and locomotive portraits throughout his career, perhaps as a way to "certify his experience"[29] with the railroad environment as well as documenting the changes he saw and endured. These two "mini" series, within the larger context of his oeuvre, further prove he was obsessively involved in making photographs that went beyond the norms of standard railroad photographic practice. They were so "different" in fact, that Everett DeGolyer Jr. wrote to Steinheimer: "your influence has me doing a number of things which are new (at least to me). Have photographed many head-ons (diesel fronts) and location signs in spots you missed."[30]

The station sign pictures appearing in the introductory pages to *Western Trains* (Golden West Publications, 1965, pages 7–9; plates 12–15 here), or the locomotive portraits in the same book (pages 28–29; plates 16–21 here; see following spread) suggest a "typological" approach to picture making—an image-gathering tactic rooted in pop art as evidenced by Andy Warhol's Campbell's Soup cans or German photographers Hilla and Bernd Becher's vernacular industrial portraits of blast furnaces or water towers—work being done at approximately the same time Steinheimer made the majority of these images. While we have no direct evidence that he saw these works, he did make occasional trips to art museums in San Francisco.[31]

Simply put: a "typology" is a collection of members of a common class or type, all photographed in a similar fashion. In this case, the station sign or locomotive is centered in the image area and the photographic frame is passive. The pictures, while mostly informational, have a quiet beauty all their own. They eschew the idea of a "masterpiece" because the photographs have more power and presence collectively than individually. At first glance the images all seem alike, but upon closer scrutiny we discern similarities and differences, and revel in the comparisons. In the case of the locomotive portraits especially, we begin to notice the variation in paint schemes, road names, types of engines, and overall design. Steinheimer, as mentioned previously, liked diesels and had no trouble committing their portraits to film—he saw diversity where others saw standardization—the faces of the locomotives taking on individual personalities. Many engines, like the Alco PA, became anthropomorphized "old friends" to the photographer (see *PA Postlude, Trains*, November 1967).

The station signs generate a slightly different response. Varied landscapes or railroad architecture hang in the background, helping to evoke a sense of place. Oftentimes these signs are the only remnants of a depot, station, or siding: they can literally signify what once was. In one three-month period (in 1966), perhaps after becoming aware of impending station closures on Southern Pacific's Coast line (south of San Francisco), Dick set about photographing every depot and siding sign from Coyote to Santa Barbara, California, making special trips to do so. These images acted as Steinheimer's personal mementos of "having been there," or perhaps as a form of visual therapy—signposts marking his own path—recording the transitory tides of life and the world of railroading he so dearly loved.

* * * * *

FIFTEEN YEARS OF FRENETIC RAILROAD photography culminated in big events occurring for Steinheimer in the early to mid-1960s. Kalmbach Publishing Company in 1963 produced a stunning monograph—that still reigns as one of the most beautifully produced books on railroads ever made—entitled *Backwoods Railroads of the West.* The book failed commercially even though it received resound-

RIDDLE

SMOKE CREEK

PORT CHICAGO

LONE PINE

PLATES 16–21
Diesel locomotive typology:
Western Pacific EMD GP35 #3029
Santa Fe EMD F7,
Burlington EMD E5 #9955,
Southern Pacific F-M H12-44 #2363,
Southern Pacific Alco #9020
Southern Pacific F-M Trainmaster #4800.

ing critical acclaim. Its creators, in retrospect, thought the book suffered from too narrow a geographic scope. But I posit its lack of success was a function of its unique content, design, and perspective—it was simply ahead of time. It was an intimate, idiosyncratic look at railroading, more artful then informational, more sublime than brawny. The average train buff hadn't graduated to a more sophisticated notion of what railroad photography could be—and wouldn't for another two decades; this core group remained mired in a Beebe aesthetic, unable to appreciate or comprehend Steinheimer's artistry (as well as the artistry of the book's design by Ted Rose). That, coupled with a narrow distribution channel for Kalmbach into the mainstream book-buying market, created a very limited audience for *Backwoods.* The price tag at the time—twenty dollars for a sheet-fed gravure book printed on premium Mohawk superfine paper—also seemed prohibitive, when other railroad books of a more generic variety were well under ten dollars. I remember seeing a dog-eared copy at a hobby store in 1970, thinking to myself (even then) how unusual and out of place its elegance seemed among the more common offerings on the shelf.

Steinheimer, while personally satisfied with *Backwoods*, nevertheless felt the sting of poor sales. He vowed in the future to do books of a more collaborative nature—showcasing the photography of others as well as his own. He questioned the stand-alone value of his work. But the feeling wouldn't last long; another important photographic milestone occurred in 1963 when Lucius Beebe included 121 of his images in *The Central Pacific and Southern Pacific Railroads.* This book, perhaps more then any previous publication or article, put Steinheimer firmly on the map. Suddenly, within the confines of a more mainstream book, people got it. His evocative photography was the light at the end of an aesthetically dark, narrowly defined tunnel.

Another turning point occurred for Steinheimer in 1965, when an exhibition of his photography entitled *The Railroad World of Richard Steinheimer* opened at the Kodak Pavilion in Grand Central Terminal in New York City and received glowing reviews in the *New York Times.* The reviewer, Jacob Deschin, extolled: "Some of his large prints . . . are breathtaking in their dramatic impact simply as visual images."[32] Clearly, he acknowledged that Steinheimer's photography went beyond "just train pictures." The exhibition ran two weeks from November 7 through November 21, later traveling in December to the George Eastman House in Rochester. In this same year, *Western Trains,* meant to be a companion piece to *Backwoods* (albeit much smaller in size) was self-published with a family loan and distributed by Golden West Publications. The book contained many images from Steinheimer's, life-altering solo trip throughout the West in the spring of 1964 (see *Done Honest & True,* Pentrex, 1999), which hit almost all the intermountain states between California and Colorado.

Even with the inclusion of this latest work, *Western Trains* received mixed reviews in the railfan press, and the critics chastised Steinheimer for making bland, motive-power pictures from other photographers its primary focus instead of his own lyrical imagery. But here too, I think the critics failed to appreciate the aesthetic advances this small volume made within the overall railroad-book market. For one thing, it had the same high production values as *Backwoods* (albeit duotone instead of gravure printing); it had a clean layout with particularly imaginative design on its introduction and chapter-opening pages; it had poetic, well-written text by Don Sims that went beyond the banal captions found in most rail books from this era. It was a gem, but one, like *Backwoods*, that seemed to struggle with what it wanted to be: art book or railroad book?

Dick, after the commercial failure of *Backwoods,* had decided to step back from the monograph concept and instead blend the work of contemporary and past master photographers (mostly from the roster and 3/4 wedgie school) into an overall unifying concept—a treatise on motive power. The book suffered because it combined the two schools of railroad photography: roster shots and lyrical ones (mostly by Steinheimer, but with some additional fine work by Hank Griffiths Jr. and Jim Erenberger) and ended up being an mish-mash, a book that lost its way, disappointing the expectations of those following Steinheimer's artistic development. Also misunderstood by the critics (but how could they have known?) is that this modest volume accurately reflected Steinheimer's predilection toward self-effacing, understated behavior. If any images had to be cut from *Western Trains,* he usually pulled his, much to the book's detriment. He simply desired to be inclusive rather than exclusive in his dealings with fellow photographers. This ability to share the spotlight, spreading the recognition, is another reason he enjoys such a devoted following. He was a kind person as well as a great photographer.

* * * * *

STEINHEIMER WAS WILDLY ENTHUSIASTIC about his job at Fairchild Semiconductor and the photography department he established there in 1962. While he enjoyed being at the center of the creative maelstrom then defining semiconductor development in America, he nonetheless longed for a chunk of time he truly could call his own. The years 1965 through 1967, based on the amount of work I saw at the DeGolyer Library, were indeed prolific ones. But he needed something more. In 1968 things began to shift and new priorities came to the forefront. Recently divorced and perhaps feeling anxious, Steinheimer approached Everett DeGolyer Jr. about purchasing his archive. DeGolyer—the son of a wealthy Texas oilman but a historian and collector of railroad negatives and photographs in his own right—

was amassing a collection on western transportation (which was eventually donated to Southern Methodist University in 1974). As of 1968 he owned over 90,000 negatives. Steinheimer, feeling the oppressive weight of his own negative collection, and worried about its safety after enduring innumerable moves over the last two decades, decided it was time to do something. Steinheimer heard of DeGolyer through fellow railfan Jim Erenberger and began a warm, increasingly personal correspondence with "Ev" that lasted ten years, until DeGolyer's premature death in 1978. The timing of this decision dovetailed nicely with other exigencies: namely his interest in pursuing railroad photography projects he'd had in mind over the last few years, and the need for money to finance them.

Chief among these was the Centennial Project. Taking a three-month leave of absence from Fairchild, from August through October 1968, Steinheimer was "venturing out on a project of my own to show the influences of the first transcontinental railroad on the present day life between San Francisco and Omaha."[33] Steinheimer's friend Chuck Fox suggests in a recent interview that Dick may have had a *National Geographic* article in mind as he was going about this project, which included two extensive car trips around the West and one flight back to Omaha. Correspondence with DeGolyer from early 1969 reveals, however, that the project was cut short for lack of funds, and due to a job offer from Fairchild in a newly established department involving television. It is noteworthy that Steinheimer's archive shows almost no evidence of black-and-white work from 1968, and in correspondence from 1969 he tells DeGolyer: ". . . except for the Centennial pictures, 1968 wasn't exactly a banner year for making new work." The Centennial Project was shot exclusively in color, and many of the images can be seen in a book Steinheimer later co-authored with his present wife, Shirley Burman, called *Whistles Across the Land* (Cedco Publishing, 1993), as well as an article appearing in the May 1993 issue of *Trains*. All the Union Pacific images from 1968 seem to be a core drilling from this unmined body of work. A lost bit of ephemera, a specially printed business card (see figure 7) found in a file at the DeGolyer Library featuring these words, "Richard Steinheimer . . . Photographing the route of the First Transcontinental Railroad for the Centennial Observance, May 1969," became the "tell" prompting this inquiry. I suffered (happily) the good fortune of a researcher unearthing a small, but substantial, nugget.

DeGolyer, as the correspondence indicates, was indeed quite enthused by the prospects of buying Steinheimer's negative collection. Steinheimer proposed a purchase price of $12,000—a dollar a negative—for work made from 1947 to 1967, and a payment schedule spanning three years was agreed upon. Subsequent deals were struck as Steinheimer sold additional negatives to DeGolyer (this time for $1.50 a piece) in 1974 and 1975 of his newest work, plus many of the earliest Baby Brownie images (made between 1945 and 1947) recently rescued by his stepfather, Dr. Robert Julian. However, it should be noted that Steinheimer, in actuality, gave DeGolyer closer to 25,000 negatives; he only charged for those he thought of interest and historic value. He included an equal number of "outtakes" in the deal. He was adamant about seeing the entire collection intact, in total, in one place. Later correspondence with DeGolyer reveals that Steinheimer (from about 1970 onward) referred to the collection as "ours," letting go of any lingering, proprietary attachment. Luckily, Dick found in DeGolyer an empathetic and ardent trustee of his life's work, as well as someone who recognized his singular artistry and passion.

In late 1968, with the first batches of negatives arriving in Dallas, DeGolyer made quick, extensive use of the collection. He was preparing an exhibition for the Amon Carter Museum in Forth Worth entitled *The Track Going Back*, a visual celebration honoring the transcontinental railroad centennial taking place in 1969. DeGolyer's sensitive text in a book of the same name accompanying the exhibition demonstrates a broad, historical understanding of the socioeconomic role railroads played in American life. It further shows insight about the simultaneous invention and development of photography and railroads from the late 1830s onward, and presages the later scholarship done by John Gruber at the Center for Railroad Photography and the curatorial staff at the Getty Museum, who mounted *Railroad Vision* in 2002, a show that explored similar themes. DeGolyer's pictorial selection reflected a seasoned eye: he championed railroad photography as a valid art form. He was also one of the first to see beyond the rolling stock and locomotives, as the book included views of stations, workers, and ephemera—railroading's mis-en-scène. He comments on two aspects, that formed the basis of his selection for the show and makes it clear why he valued Steinheimer's work:

> Two major efforts have been made in the selection for the show and for this catalog. The first is to show the railroad in its environment. On the western transcontinentals, rugged topography is the dominant characteristic and an effort has been made to emphasis this all-important factor. Another environmental factor of importance is climate. Railroads are truly twenty-four-hour-a-day, all weather operations of astounding reliability, and this is no where better exemplified than in those operations of the roads that extend singly, or in combination, from the Missouri and Mississippi Rivers to the Pacific Coast.

He enthusiastically included forty-two Steinheimer images in the catalog, the largest number by any photographer, and the show hung from May 8 through July 4, 1969, at the Amon Carter. It would be Steinheimer's first inclusion in a major museum exhibition. Although DeGolyer elliptically mentions the book's (and show's) preparation in correspondence to Steinheimer in 1968, it appears Dick overlooked

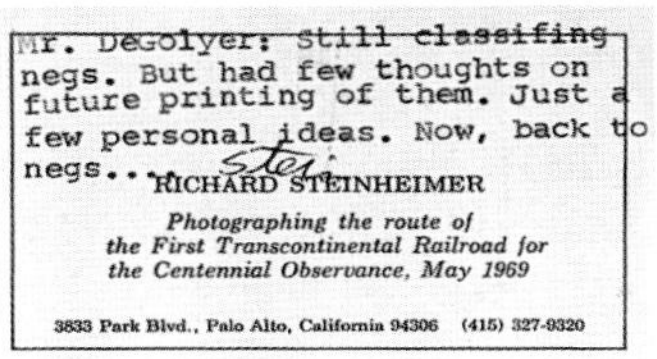

Mr. DeGolyer: Still classifing negs. But had few thoughts on future printing of them. Just a few personal ideas. Now, back to negs...
RICHARD STEINHEIMER
*Photographing the route of the First Transcontinental Railroad for the Centennial Observance, May 1969*
3833 Park Blvd., Palo Alto, California 94306 (415) 327-9320

FIGURE 7
Steinheimer's business card for his Centennial Project from 1968, courtesy DeGolyer Library, Southern Methodist University, Dallas, Texas.

his comments and was unaware of both until 1974, when he mentions seeing the book (for the first time) in Montana at the home of Adam Gratz, an engineer on the Milwaukee Road he often photographed.[34]

Other projects under consideration in the 1970s included a *Winter in the West* book with Howell-North of Berkeley. While never coming to fruition in hardcover format it did become an extended *Trains* magazine photo-essay in February 1974 entitled *California in the Winter.* He also contemplated doing a book about air pollution while living near Lake Arrowhead above Los Angeles, but nothing came of the idea.[35] The other major project that consumed Dick during this time was a book on the Milwaukee Road, a railroad he'd photographed for twenty-one years. The DeGolyer-Steinheimer correspondence reveals that Dick first broached the idea of such a book in September 1973, pitching the project to DeGolyer as

PLATE 22
Union Pacific station platform and milk cans, Cascade, Idaho, 1964.

> a kind of three-way venture between you (with your museum collection) and R.V. Nixon and myself . . . to do with the Milwaukee Road electrification on the eve of its demise. Some of my old shots are pretty good, and you should see my new ones. It could be a "Milwaukee Road Electrification Pictorial" or "Electric Days on the Milwaukee Road." Combined with help from the Milwaukee (Road) PR department in Chicago, we could do something which . . . would be a feather in any publisher's hat.

While DeGolyer enthusiastically embraced the project, he couldn't fund its publication or help Dick pursue a publisher. Steinheimer eventually enlisted the help of Richard F. Lind, another photographer known for his skillful imagery. The two men, however, ran into creative differences during the fall of 1975 over how the material should be handled and parted company.[36] The book had a protracted gestation period lasting seven years before seeing publication. It involved a lot of legwork for Steinheimer as he gathered art from disparate corners of the country. Northwest photographers R. V. Nixon and Wade Stevenson made significant contributions; old friend Wallace Abbey aided the search as the newly appointed communications officer for the Milwaukee Road;[37] work from gifted younger photographers like Ted Benson and Max Tschumi also graced its pages. Extant letters documenting correspondence between Steinheimer and General Electric seeking essential historic photographs show how far Dick cast his research net. He was indefatigable in his desire to tell the complete story. Finally, *The Electric Way Across the Mountains* (Carbarn Press) came out in 1980 to enthusiastic praise. It became a lasting, affectionate tribute to the Milwaukee Road in general, and the men who ran the railroad in particular, and it was Steinheimer's favorite among the books he'd produced.

* * * * *

IN AUGUST OF 2002 MY WIFE and I had the good fortune of viewing the files from those years—about two thirds of the Steinheimer images in the DeGolyer Collection at SMU—spending four days "white-gloving" negatives from their envelopes. From a research standpoint it was enlightening to see a cross-section of one artist's output and its evolution over a thirty-year period. The amount of work was overwhelming, wide-ranging, and obsessive. When did Dick have time to do anything else? I knew that Steinheimer was prolific, but I wasn't prepared for how prolific. Fascinating, too, were the "chatty" notes typed by Steinheimer outside the 4 x 5 envelopes, detailing aspects of trips, locations, stories, fragments of thought. He felt compelled to give Everett and future researchers as much information as possible regarding his odyssey as a railroad photographer. The notes were typed as Steinheimer prepared to transfer the remaining negatives to DeGolyer in October 1969—a task which took close to 1,000 hours—as mentioned in a telegram sent to DeGolyer as Dick boarded *The City of San Francisco* headed for Dallas (via Denver), with the rest of his collection in tow. It seemed fitting to ride a favorite passenger train to complete this transaction.

Along with the stories these envelopes contained, there were also detailed printing instructions for darkroom technicians. They confirm that Steinheimer was a master printer attentive to nuances in contrast; he was also very meticulous about selective burning and dodging (darkening and lightening certain areas of a print with either more or less exposure), bordering on perfectionism. His specific instructions were most evident for blizzard pictures, where he suggested printing with "a hard #4 or #5 paper to enhance the properties of snowfall" (see plates 22, 37, and 98). After reviewing many original Steinheimer prints (and seeing variations of several) I was struck by how much his newspaper/photojournalism background influenced his printing style. He oftentimes (and in subtle, almost unnoticeable ways) burned and dodged to "knock down" a distracting highlight here, deepen a shadow there, all to bolster the photograph's impact, which is a key component of journalistic work. But as great a printer as he was, there's a subtle irony in his machinations: on some level his images needed no additional help from darkroom wizardry. The impact of his photography was assured at the shutter's release.

The 1980s were also a fertile period for Steinheimer—a decade that saw major transformations in his work. In 1983 he switched to color, eschewing black and white completely. Shooting color slides (for most railfans) in the 70s and 80s had become infinitely easier since Kodachrome's early days—a film introduced into the marketplace around 1935. Contemporary slide films were heads and shoulders above earlier transparency materials: they were faster, had better latitude (a drawback with the original Kodachrome), and were archivally more stable. Slides, too, were less cumbersome to store

ATLANTIC
COAST
LINE

than prints, and required less handling to view the finished product. There was no cumbersome darkroom work to do, and slides provided an instantaneity in viewing results. It was a medium suited to the accelerated pace of the times.

The home slide show, like watching television or a film, also provided a radically different viewing experience then looking at a two-dimensional 8 x 10 print—it was an activity better suited to gatherings and group functions too. This notion fueled the creation of national slide extravaganzas like California's *Winterail*. These events encouraged color slide photography and became venues for elaborate audiovisual presentations. Steinheimer took part in this trend with three major shows at *Winterail* in 1983 (on Amtrak), 1988 (a collaboration on A. J. Russell with wife Shirley Burman), and 2000 (a fifty-year retrospective).

This shift to color also coincided with changes in the railfan press and railroading itself. *Trains* Magazine begun running color covers full-time in March 1972, with an occasional four-color center spread. But it wouldn't be until the development of laser-scanning technology during the early 80s, and its growing prevalence in the printing industry (and attendant cost savings), that widespread use of color in rail magazines occurred. *Trains* went predominately four-color in 1988. Aiding this eventuality was a maturing readership familiarized with color television, sophisticated advertising, and—on the railroading front—locomotives and rolling stock displaying vibrant paint schemes. They demanded color because they were used to it. No more boxcar brown or engine black, the world looked better captured on Kodachrome, as Paul Simon's pop song from 1973 portended. Steinheimer drove toward this rainbow ready to exploit available opportunities. By switching to color he refreshed and rechanneled his creative energies.

The mid- to late 80s also signaled another paradigm shift for *Trains*, *Pacific Rail News*, and *Locomotive and Railway Preservation* in terms of picture content. Editors and art directors began publishing more adventuresome photography with *Trains Illustrated*, premiering in 1988, representing the fullest expression of this new direction. Photographers of this generation, used to more sophisticated visuals featuring jump cuts (the MTV-ification of film and television), uncommon camera angles, daring compositions, or dramatic lighting were even more prone to experimentation than their predecessors. They took pictorial cues from the past but recast them in a contemporary way, the photographic torch being passed again as it had from Beebe to Steinheimer and his contemporaries. And like Beebe, Steinheimer established groundbreaking precedents in the field that begged to be emulated, furthered, and finally set aside, his images a database of archetypes to fall back on and advance from. Undeniably, photographic headway had been made since the days of Russell and Jukes, and Beebe's first publications in the 1930s, but Steinheimer proved to be one of the preeminent catalysts sparking this evolution in the later half of the twentieth century.

In 1983 he was given a Lifetime Achievement Award by the Locomotive and Railway Historical Society for his photography; 1982 and 1983 saw the publication of two popular books called *Growing Up With Trains* (a Northern and Southern California version); an all-color book called *Diesels Over Donner* hit the shelves in 1989, followed by the aforementioned co-publication with Shirley, *Whistles Across The Land* (1993). *Done Honest & True* (from 1999) contained an epic account of his "on-the-ground" travels around the western United States pursuing trains and railroads over five decades. This volume, a captivatingly written swan song about Steinheimer's efforts by renowned photojournalist and artist Ted Benson, formed a wonderful bookend to Dick's fifty years of railroad photography.

* * * * *

IN THE FALL OF 1990 STEINHEIMER was able to realize a two decades-old dream: to meet Naotaka Hirota, Japan's greatest living railroad photographer and someone Dick had admired since 1957. From the first photographs of Hirota's he'd seen in *Trains*—and especially a special six-page tribute to the Raven Black C-62 by Hirota in a September 1970 issue of the same magazine—Steinheimer felt an affinity for Japanese railroad photographers, whose work was stylistically similar to methods he employed. Steinheimer often jokingly referred to this brand of picture taking as "kamikaze" photography: a term not meant derisively but one suggestive of an all-out, "do-or-die" approach, which the Japanese specialized in and Dick deeply admired. Steinheimer, too, appreciated the emotionality, lyricism, and reverence for nature found in Japanese rail photography (see page 35 *Trains,* March 1959), not to mention the daring photo selections the magazines' editors made. Steinheimer also found the bold graphic presentation of the material and the superb production values compelling. Publications like *Japan Railfan Magazine* placed as much emphasis on the *act* of photography as they did on the railroad or train content depicted—a significant departure from most American railfan magazines prior to 1980. In fact, a publisher had approached Steinheimer in 1989 about a possible book project for Japan, so Dick knew railfans in the Far East admired what he did. Japanese railfans especially enjoyed articles about American railroading when they contained Steinheimer photographs. He was a kindred spirit.

Kentaro Hirai and Ed Delvers arranged the meeting between Steinheimer and Hirota in spring 1990; travel itineraries were finalized for the following October. Over the course of a week, the two traveled (with a small supporting entourage) throughout California, first visiting favorite Steinheimer locations on Donner Pass and then driving south to see the Tehachapi Loop. In the evenings, after the cameras were put away, extensive interviews ensued in hotel rooms, where Delvers (who speaks fluent Japanese) translated the conversation and questions between these two giants of rail photography. When asked what makes a great train photographer, Hirota's comments (which in some ways had nothing to do with photography; it's interesting to note his uncle is a well-known Zen monk) can best be summed up by words published in his own debut book *The Lure of Japan's Railways* (The Japan Times, Ltd, 1969):

> From my own valuable experience, I've learned that you have to have strong legs to run about in the mountains on location, and you have to have the swing of a baseball batter when you shoot a train whizzing past. I have realized anew that you need a sound body as well as a sound mind to make a sound judgement in seizing a 1/1000 second shutter chance.

This description sounds deceptively simple, like a Zen koan ("when we carry water, we carry water; when we chop wood, we chop wood"), but this economy of expression belies a broad reservoir of creativity. Steinheimer felt close to Hirota's essence and appreciated the meditative quality of his work. The two obviously shared a similar belief system and desire to push themselves beyond known limits. Both relied on a sturdy constitution, physical strength, stamina, and agility; both exhibited a no-nonsense, not-too-much-thinking, "just-do-it" approach—practicing Nike's slogan long before the phrase became mainstream. On location shooting trains at Tehachapi Loop, Hirota—small in height and slim in build—had a quiet calmness, often staying in one location (camera on tripod) making very deliberate compositions as several trains went by. Steinheimer (at this point shooting strictly 35mm with long lenses) had a more fluid, grab-bag interaction with the train, environment, and light. But then, conversely, at another point Stein might be on a tripod and Hirota would be handholding his Canons, reacting to where the sun was, and calculating exposures accordingly. In other words, both photographers had similar modi operandi and were adapting freely based on stimuli confronting them: about as "Zen" as it gets. There were many paths to the photographic mountaintop and neither had any compunction in deviating to find a way there. There were always ample "shutter chances" no matter what path was chosen.

The story about their exploits was published in Japan's *Train* magazine (issue 193) in January 1991, featuring their joint portrait on the cover with twenty-two pages of photos and text inside (see figures 8 and 9) It was a watershed event for Japanese publishers, who had long admired Steinheimer's work through early issues of *Trains* magazine.

* * * * *

THINGS HAD COME FULL CIRCLE. As my wife and I stood in Dick's studio in July 2002, behind his and Shirley's modest home in Sacramento's Curtis Park section, the clutter of genius that I first encountered in Palo Alto almost thirty years earlier surrounded us. Dick at seventy-five stays engaged; his friendly dishevelment is still a source of humor—he's a walking work-in-progress. He often looks like a 6′6″ (his actual height) Einstein and probably exhibits (I imagine) the same affable quirkiness and flashes of brilliance the physicist did. As we poured through those same print boxes, turning his studio inside out, leaving nothing unopened or uninspected, he would sidle up to the table and tell stories about each image. His memory astounded us: the names, places, and dates all came back easily. As he stood over our shoulders, shyly pointing out details, we could sense his abiding affection for the past he experienced and photographed. Standing there he appeared heroic, like so many of the photographs before us.

I've often thought it important, no matter what age, to have someone to emulate and look up to. When I was a kid of eighteen Steinheimer became that all-important first hero for me and remains so today. Even though we had only infrequent contact over the past several decades, he was a mentor in absentia. His published photographs came to feel like messages of encouragement, messages that said, "carry on, you'll get there." His persistent passion for accomplishment and work well done acted as a beacon—you could see it right there on the page. He probably never knew this and possibly wouldn't have cared. But that's the invisible positive force he's had on us. And make no mistake: Dick has had a profound influence on two generations of railroad photographers in the United States—first for the baby boomers who immediately followed his generation; and then later the Gen-X photographers who sought out his publications to find out

FIGURE 8
Steinheimer and Hirota discuss photography trackside, near Mojave, California, 1990. Photograph by Jeff Brouws.

FIGURE 9
Cover of Japan's *Train* magazine featuring portrait and story on historic Steinheimer/Hirota meeting. Photo taken at Caliente, California, 1990. Courtesy *Train* magazine. Photograph by Jeff Brouws.

who this "Stein" guy was, and just what made his work so important.

Unlike most celebrities in American life that fall victim to the "shooting-star" syndrome—which sends stars up and then spits them out once they pass their zenith—Steinheimer has never disappointed or faded away. He maintains a humility, graciousness, and integrity that transcend fashion: he's the real deal. Fifty years of hard work that saw every available dollar go back into his photography—fifty years of photography that created a lot of self-satisfaction but little remuneration—prove the point. He established his own criteria for success. He went about his business, making images that always remained fresh and providing unfailing support to others as a mentor and friend. These are the things that last. While countless rail photographers have reverentially imitated his work, or used it as a basis from which to extend their own explorations, the greatest gift Steinheimer passed on to younger photographers (besides the visually graphic ones) is the urgent desire to shoot trains in a novel way. He encouraged others, by example, to blaze trails off the ones already well trod. A fellow photographer who traveled with Steinheimer recently stated: "Dick was always a bit nebulous in discussing photo technique or philosophy—you could learn more by watching him, absorbing as if through osmosis. His enthusiasm for what he was doing was the real teacher."

Notes

1 "Edward Hopper's Objects," letter from Hopper to the editor, Nathaniel Pousette-Dart, *The Art of Today* 6 (February 1935), 11.
2 Phone interview with Jane Thompkins (Steinheimer's sister), August 2002.
3 Phone interview with Jane Thompkins, August 2002.
4 Interview with Richard Steinheimer, Sacramento, California, July 2002.
5 Robert Taft, *Photography and the American Scene* (New York: Dover Publications, 1938), 309–10.
6 Naomi Wolf, *World History of Photography* (New York: Abbeville Press, 3rd ed., 1997), 447.
7 Lucius Beebe, *Great Railroad Photographs USA* (Berkeley, California: Howell-North Books 1964), 84.
8 George Eastman House website, Timeline of Photography: www.eastman.org/5_time line/5_index.html.
9 Letter from Cornelius Hauck to John Gruber, 1999, and interview with Cornelius Hauck, November 2002.
10 "A Tribute to Fred Jukes, Boomer Railroad Photographer," *Colorado Railroad Annual #9* (Colorado Railroad Museum, Golden, Colorado, 1971), 6.
11 Lucius Beebe and Charles Clegg, *The Age of Steam* (Berkeley, California: Howell-North Books, 1972), 263.
12 Interview with Cornelius Hauck, November 2002.
13 Everett DeGolyer Jr, *The Track Going Back* (Forth Worth, Texas: Amon Carter Museum 1969), 6.
14 It should be noted the Denver Public Library houses a fine selection of work by seminal railroad photographers working in Colorado in the nineteenth and twentieth centuries: W. H. Jackson, L.C. McClure, Jackson Thode, Richard Kindig, Otto Perry and Fred Jukes. All photographers were early pioneers in action photography; and all, except Jackson and Jukes, had no commercial outlet for their work. They were enthusiasts enamored with rail-roading, and all labored at other professions.
15 Linn Westcott, "Railroad Photography," *Trains*, December 1945, 21.
16 Frederick H. Richardson and F. Nelson Blount, *Along the Iron Trail* (Rutland, Vermont: Tuttle Publishing Company, 1938), 192.
17 Linn Westcott, "Railroad Photography," *Trains*, December 1945, 20.
18 Andria Daley Taylor, "Authors Celebrate Glory of Steam Era," *Vintage Rails*, Fall 1997, 42.
19 J. Chislain Lootens, *Lootens on Photographic Enlarging and Print Quality* (Baltimore, Maryland: Fleet-McGinley, Inc, 1949), v, 177.
20 Andria Daley Taylor, "Authors Celebrate Glory of Steam Era," *Vintage Rails*, Fall 1997, 94.
21 Linn Westcott, "Railroad Photography."
22 Goodrich, Lloyd, *Edward Hopper* (New York: Abrams, 1971), 164.
23 William Kittredge, *Perpetual Mirage: Photographic Narratives of the Desert West* (New York: Whitney Museum of American Art, 1996), 63–64.
24 Richard Steinheimer, *Backwoods Railroads of the West* (Milwaukee, Wisconsin: Kalmbach 1963), 174.
25 Steinheimer interview with Carbarn Press, October 20, 1980.
26 Interview with Persia Wooley, October 2002.
27 Susan Danly and Leo Marx, *The Railroad in American Art* (Cambridge, Massachusetts: MIT Press, 1988), 101.
28 Lucius Beebe, *Great Railroad Photographs USA*, 162.
29 Susan Sontag, *On Photography*, (New York: Farrar, Straus & Giroux 1973), 9.
30 Letter from DeGolyer to Steinheimer, December 3, 1975.
31 Interview with Persia Wooley, October 2002.
32 Jacob Deschin, "The Camera as Aid To Hobbies," *New York Times*, November 7, 1965.
33 Letter from Steinheimer to DeGolyer, July 31, 1968.
34 Letter from Steinheimer to DeGolyer, January 21, 1974.
35 Letter from Steinheimer to DeGolyer, September 10, 1973.
36 Letter from Steinheimer to DeGolyer, November 26, 1975.
37 Letter from Steinheimer to DeGolyer, May 24, 1977.

# Plates

PLATE 23 Western Pacific westbound freight with F7 #919D, as seen through windshield of Land Rover, Smoke Creek, Nevada, 1965

PLATE 24 Denver and Rio Grande Western symbol freight SFO at Arena, Colorado, 1964

PLATE 25 Union Pacific freight train with DD40Xs, Echo Canyon, Utah, early 1970s

PLATE 26 Midland Continental freight train, Wimbledon, North Dakota, 1957

PLATE 27
Great Northern brakeman signals with flare, Gold Bar, Washington, 1964

PLATE 28
Chicago, Burlington and Quincy depot, Powder River, Wyoming, 1964

PLATE 29 Milwaukee Road Little Joes switch cars at Sappington Hereford Ranch, Willow Creek, Montana, 1964

PLATE 30 Northwestern Pacific *Redwood* passenger train, Scotia, California, 1957

PLATE 31
Milwaukee Road freight train on trestle, near Pee Dee, Idaho, 1972

PLATE 32
Southern Pacific freight with Fs, Palisade, Nevada, 1962

SOUTHERN PACIFIC
C&NW
Burlington

PLATE 33 Union Pacific "Big Blow" Turbine X-20 and freight train, Cheyenne, Wyoming, 1964

PLATE 34 Chicago, Burlington and Quincy train #29, Wind River Canyon, near Thermopolis, Wyoming, 1964

PLATE 35 Southern Pacific commuter looking out window of Harriman Coach, Palo Alto, California, 1955

PLATE 36 Milwaukee Road, welder and boxcabs, Cedar Falls, Washington, 1967

PLATE 37 Colorado and Southern 2-10-2 #900 on turntable, Cheyenne, Wyoming, 1956

PLATE 38 Southern Pacific freight at Cantara Loop near Dunsmuir, California, 1964

PLATE 39 Great Northern *Empire Builder* departs King Street Station, Seattle, Washington, 1964

PLATE 40 Southern Pacific freight on Donner Pass, near Cisco, California, 1972

**PLATE 41**
Northern Pacific FT #5402A with
freight train, Pasco, Washington, 1965

**PLATE 42**
Southern Pacific cab-forwards
at Taylor Yard Roundhouse,
Los Angeles, California, circa 1949

PLATE 43
Southern Pacific steam engines
being scrapped with 2-10-2 in background,
Richmond, California, 1956

3659
SOUTHERN PACIFIC
3659

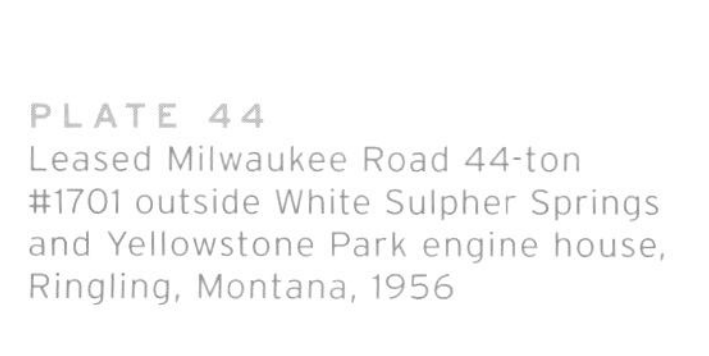

PLATE 44
Leased Milwaukee Road 44-ton #1701 outside White Sulpher Springs and Yellowstone Park engine house, Ringling, Montana, 1956

PLATE 45
Longview, Portland and Northern freight train parked at crossing, Battleground, Washington, 1963

NP 6521
NORTHERN PACIFIC
RAILWAY
LP & N
112
RAIL
CROSSING
ROAD
RAIL
CROSSING
ROAD

PLATE 46
Southern Pacific, lone worker and roundhouse, San Luis Obispo, California, mid-1950s

2836
SOUTHE
SOUTHE

PLATE 47 Northern Paciifc 0-6-0 #1110 switching cars, Chehalis, Washington, 1956

PLATE 48 Southern Pacific freight yard in snow, Dunsmuir, California, 1959

PLATE 49 Colorado and Southern train #8-28 with E5 #9955, Pueblo, Colorado, 1964

PLATE 50 Stockton, Terminal and Eastern engine house, Stockton, California, 1970

PLATE 51 Atchison, Topeka and Santa Fe westbound *El Capitan* at dusk, Yampai, Arizona, 1968

PLATE 52 Western Pacific's *California Zephyr* at Fremont, California, 1970

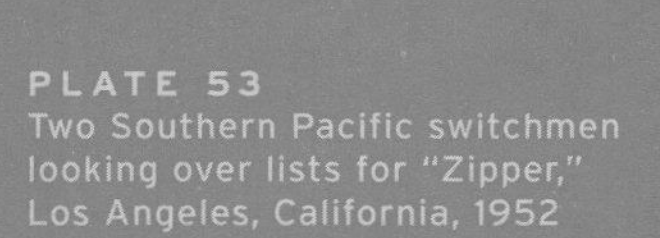
**PLATE 53**
Two Southern Pacific switchmen looking over lists for "Zipper," Los Angeles, California, 1952

**PLATE 54**
Colorado and Southern 2-10-2 #900 getting ready to leave roundhouse, Cheyenne, Wyoming, 1957

900

PLATE 55 Oregon and Northwestern Baldwin AS616 locomotive and log train, near Burns, Oregon, 1960s

PLATE 56 San Francisco State Belt Railway Alco switcher being repainted, San Francisco, California, 1958

PLATE 57
Southern Pacific freight,
as seen from State Highway 99,
Cottage Grove, Oregon, 1967

MID-AMERICA
SP

**PLATE 58**
San Diego & Arizona Eastern yard,
San Diego, California, 1952

**PLATE 59**
Northern Pacific freight #603,
Jamestown, North Dakota, 1957

6008A
E.W. 9-4
E.H. 12-5

PLATE 60 Denver and Rio Grande Western freight train, west of Antonito, Colorado, 1950s

THOMPSON-NICOLA REGIONAL DISTRICT LIBRARY SYSTEM

PLATE 61
Mount Hood Railroad,
pacing shot of Alco switcher,
Dee, Oregon, 1963

50
MOUNT HOOD RAILROAD
50

PLATE 62 Northern Pacific FT #5402A with freight train enters yard, Pasco, Washington, 1965

PLATE 63 Chicago, Burlington and Quincy daily freight, Parkman, Wyoming, 1953

Cushioned
Ride
278 A
GREA

PLATE 64
Great Northern F7 #278A
coupling to train,
Seattle, Washington, 1964

PLATE 65
San Diego and Arizona Eastern
freight on Goat Canyon trestle,
Carriso Gorge, California, 1953

PLATE 66
Wheel and journal
box of a PFE Reefer,
Roseville, California, 1965

PLATE 67
Midland Continental track
laborers at grade crossing,
Jamestown, North Dakota, 1957

RAILROAD
CRO SING

PLATE 68 Bridge and Building foreman inspects track #1 on Southern Pacific main line, east of Donner Summit, California, 1965

PLATE 69 Colorado and Southern steam locomotive at stockyard, Horse Creek, Wyoming, 1956

PLATE 70 Southern Pacific's *City of San Francisco*, Applegate, California, 1961–62

PLATE 71 Milwaukee Road freight on branch line near Manhattan, Montana, 1964

PLATE 72 Chicago, Burlington and Quincy freight train #75 in Wind River Canyon, near Thermopolis, Wyoming, 1964

PLATE 73 Union Pacific's *Los Angeles Limited* at Sullivan's Curve, Cajon Pass, California, 1952

PLATE 74
Dan Murray, Milwaukee Road conductor, Malden, Washington, 1964

PLATE 75
Southern Pacific locomotive shops, Sacramento, California, 1956

3643
744
STORES

PLATE 76 Missouri Pacific's southbound *Colorado Eagle*, Palmer Lake, Colorado, 1964

PLATE 77 Southern Pacific F7 #6229 with time freight, Antelope, California, early 1960s

PLATE 78 Southern Pacific's Glendale Tower, Glendale, California, 1950

PLATE 79 Western Pacific symbol SWG freight with Fs, Stockton, California, 1965

PLATE 80 Great Northern westbound freight train with F7 #460A, near Browning, Montana, 1964

PLATE 81 Southern Pacific freight train at Crystal Lake snowsheds, near Cisco, California, 1967

PLATE 82 Dutch Brantz, SD&AE roundhouse foreman, Calexico, California, 1953

PLATE 83 Southern Pacific 0-6-0 switch engine #217, Oakland, California, 1956

PLATE 84 Carbon County Railway SW1200s and empties, near Columbia, Utah, 1960s

PLATE 85 Tucson, Cornelia and Gila Bend Railway freight train, Black Gap, Arizona, 1952

O.S.L. 47907D
UNION PACIFIC
LIVESTOCK DISPATCH
O.S.L.
47907D
UNION PACIFIC

**PLATE 86**
Union Pacific, brakeman
on top of cattle car,
Cajon Pass, California, early 1950s

**PLATE 87**
Denver and Rio Grande Western
depot and train,
Spanish Fork, Utah, 1964

PLATE 88 Southern Pacific & Western Pacific crossing, Flanigan, Nevada, 1961

PLATE 89 Chicago, Burlington and Quincy northbound freight with SD9 #448 in Sheep Canyon, near Greybull, Wyoming, 1964

PLATE 90
Southern San Luis Valley Railroad
caboose at sunset, Blanca, Colorado, 1961

PLATE 91 Southern Pacific's *Del Monte* passenger train with GP9 #5600, San Francisco, California, 1964

PLATE 92 Atchison, Topeka and Santa Fe Engine Terminal, Chicago, Illinois, 1965

PLATE 93 Southern Pacific 4-6-2 #2484 and freight train, Oakland, California, 1956

PLATE 94 Union Pacific freight train with DD35 #71, Cima, California, 1977

PLATE 95 Denver and Rio Grande Western freight train near Provo, Utah, 1964

PLATE 96 Southern Pacific troop train with GS-4, near Casmalia, California, 1955

PLATE 97 Atchison, Topeka and Santa Fe freight train near Monolith, California, early 1970s

PLATE 98 Denver and Rio Grande Western freight exiting Moffat Tunnel, East Portal, Colorado, 1964

PLATE 99
Southern Pacific's
*City of San Francisco* arrives at
Oakland, California, 1962

PLATE 100
Southern Pacific,
hostler and Alco #9018 at diesel shops,
Roseville, California, 1966

**PLATE 101**
Northern Pacific fireman
taking on water for 2-8-2 #1708,
Auburn, Washington, 1956

**PLATE 102**
Midland Continental freight train,
Jamestown, North Dakota, 1957

**PLATE 103**
Southern Pacific eastbound freight
with 2-10-2 #3652 leaves Colton Yard,
Colton, California, 1949

X3652

FRUIT EXP

P.F.E.

6758

CAPY 80000

LD LMT 83000

LT WT 53000 NEW 11-

3652

NP

PACIFIC

**PLATE 104**
Union Pacific freight train with F7
#1433, Horseshoe Bend, Idaho, 1964

**PLATE 105**
Southern Pacific,
Man and wigwag signal at crossing,
Lodi, California, 1971

PLATE 106
Union Pacific's *City of Los Angeles*,
Green River, Wyoming, 1952

PLATE 107
Southern Pacific, steam
and diesel freight action,
Port Costa, California, mid-1950s

N&W
NORFOLK
AND
WESTERN
20 354
1336
SOUTHERN PACIFIC

PLATE 108
Southern Pacific Harriman coaches
at 3rd and Townsend depot,
San Francisco, California, 1965

MEN
AT
WORK

PLATE 109 Union Pacific Gas Turbine #54 and freight, Wahsatch, Utah, 1953

PLATE 110 Great Northern's *Empire Builder*, East Glacier, Montana, 1964

PLATE 111 Union Pacific freight train with reefers, Ogden, Utah, 1952

PLATE 112 Northern Pacific laborer and RS-1 locomotive, Duluth, Minnesota, 1957

PLATE 113 Denver and Rio Grande Western trainman drains cylinder cocks, Durango, Colorado, 1961

PLATE 114 Southern Pacific 2-10-2 #3625 and trainmen, Saugus, California, 1948

PLATE 115 Southern Pacific's *San Joaquin Daylight* at tunnel #3, Tehachapi, California, 1952

PLATE 116 Cab ride on the Oakland Terminal Railroad, Oakland, California, 1955

PLATE 117
Denver and Rio Grande Western freight, Cumbres, New Mexico, 1961

PLATE 118
Milwaukee Road, moonlit portrait of Little Joe #E77, near Avery, Idaho, 1973

PLATE 119 Chicago, Burlington and Quincy freight yard, East St. Louis, Illinois, 1965

PLATE 120 Denver and Rio Grande Western's *California Zephyr*, near Plainview, Colorado, 1964

PLATE 121 Southern Pacific, turntable and roundhouse with F3 #6164 and Krauss-Maffei #9010, Bakersfield, California, 1964

PLATE 122 Hobo sleeping near Southern Pacific yards, Roseville, California, 1964

PLATE 123 Sierra Railroad railfan trip as seen from car, Northern California, late 1950s

PLATE 124 Southern Pacific's *City of San Francisco*, near Wells, Nevada, 1965

NORTHERN PACIFIC
NP
6096

PLATE 125
Southern Pacific
westbound freight at
Mt. Hebron, California, 1960s

PLATE 126
Denver and Rio Grande Western
engineman comes off duty,
Salt Lake City, Utah, 1952

1406
1406
Rio Grande
1205
115
D&RGW

PLATE 127 Southern Pacific's *City of San Francisco* in Cold Stream Canyon, near Andover, California, 1962

PLATE 128 Great Northern's *Empire Builder*, near Browning, Montana, 1964

E75

PLATE 129
Milwaukee Road Little Joe and freight, near Haugen, Montana, 1960s

PLATE 130
Denver and Rio Grande Western, Chama, New Mexico, 1961

**PLATE 131**
Southern Pacific,
train order signal,
Gilroy, California, 1966

**PLATE 132**
Southern Pacific eastbound
freight on Suisun Bay Bridge,
Benicia, California, early 1950s

X4150
X4150
4150

PLATE 133
Union Pacific's
*North Platte Valley Express*,
Yoder, Wyoming, 1953

PLATE 134 Colorado and Southern northbound freight, above Horse Creek, Wyoming, 1956

PLATE 135 Denver and Rio Grande Western meet between two freight trains, Price Canyon, Utah, 1951

PLATE 136 Terminal Railroad Association of St. Louis switcher on bridge, East St. Louis, Illinois, 1965

PLATE 137 Southern Pacific, cab-forward #4363 and sectionman on speeder, Saugus, California, early 1950s

PLATE 138 Southern Pacific freight with U25B #6765, near Dos Cabezas, California, 1967

PLATE 139 Union Pacific's *City of San Francisco*, Green River, Wyoming, 1952

PLATE 140 Southern Pacific's *Lone Pine Local*, Cantil, California, 1971

PLATE 141 Southern Pacific commute train with 4-8-4 #4405, South San Francisco, California, mid-1950s

PLATE 142
Southern Pacific,
laborer washing cab windows
on *City of San Francisco*,
Oakland, California, 1966

PLATE 143
Southern Pacific,
Taylor Yard Diesel Facility,
Los Angeles, California, 1949

6004
SOUTHERN
LINES
PACIFIC
SMOKING
- POSITIVELY -
PROHIBITED
ABOARD DIESEL LOCOS.
OR ON UPPER & LOWER
WORKING LEVELS

PLATE 144
Atchison, Topeka and Santa Fe depot interior and operator's desk, Palmer Lake, Colorado, 1964

Vincent's
Complete
INSURANCE & BONDING SERVICE
MELROSE
4-5584
30

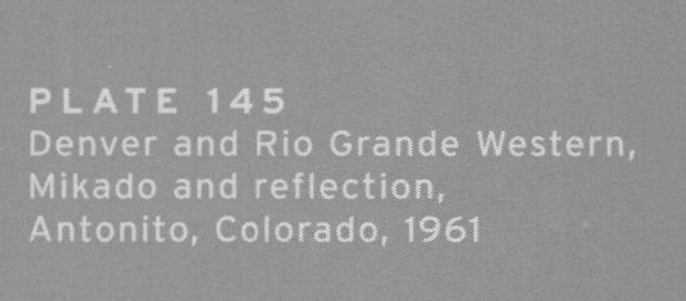

**PLATE 145**
Denver and Rio Grande Western,
Mikado and reflection,
Antonito, Colorado, 1961

**PLATE 146**
Denver and Rio Grande Western,
Mikados and freight,
near Falfa, Colorado, 1961

D&RGW
06217
WATER CAR
Rio Grande
494
Rio Grande
483

PLATE 147 Denver and Rio Grande Western 2-10-2 #1403 with coal drag, Thistle, Utah, 1951

SANTA FE
SAN

**PLATE 148**
Atchison, Topeka and Santa Fe *San Diegan* prior to departure, San Diego, California, 1952

**PLATE 149**
Abandoned Oakland Mole Train shed, Oakland, California, 1962

PLATE 150
Southern Pacific,
disabled F7 units on turntable,
Norden, California, 1967

PLATE 151
Southern Pacific
Burbank Junction Tower,
Burbank, California, 1950

PLATE 152
Southern Pacific
freight with SD45 #8845,
Lodi, California, 1971

PLATE 153
Southern Pacific
F-M Trainmaster #4813
and commute train,
Palo Alto, California, 1964

SOUTHERN PACIFIC
4813

# Captions

**PLATE 1** Colorado & Southern engine house with SD9 #828, Leadville, Colorado, 1968.

The summer of 1968 finds a railroad worker opening creaky wooden doors of Colorado and Southern's engine house in Leadville shedding light on its darkened interior while simultaneously making visible the sun-washed face of SD9 #828. The SD9's sole raison d'être is the 14-mile spur to the Climax molybdenum mine atop Fremont Pass. There, at 11,300 feet, main-line railroading reached its highest point in the continental United States. This remote branch, which featured steam locomotives until 1962, also carried the fabled trains of the Denver, South Park and Pacific in an earlier part of the twentieth century.

**PLATE 2** Father and son watching the Southern Pacific's *West Coast* depart, Glendale, California, 1949.

In what must have been an early important symbolic image for Steinheimer, a father and son watch the late-night rail activity at Glendale, California, as train #59, the SP's *West Coast,* prepares to depart. Throughout his entire body of work, Steinheimer repeatedly photographed lone, or semi-alone figures, within the railroad landscape. In the DeGolyer files there are negatives of his son Alan peering into a passenger train window, perhaps echoing Steinheimer's own father/son/depot experience in 1935, as well as pictures of Steinheimer's stepfather, Dr. Robert Julian, alone on the platform in Burbank in the 1960s, saying good-bye to Dick, who's leaving on a passenger train to return to the Bay Area.

**PLATE 3** Union Pacific freight train at Summit, Cajon Pass, California, April 1946.

One of Steinheimer's earliest images pays homage to Lucius Beebe and the 3/4 wedge aesthetic then prevalent in American railroad photography, as a Union Pacific freight led by a husky 2-10-2 blows through Summit in California's Cajon Pass.

**PLATE 4** Denver and Rio Grande Western freight crossing State Highway 72, Coal Creek Canyon, Colorado, 1964.

Four EMD GP30s lead a westbound D&RGW freight across the bridge spanning Colorado State Highway 72 near the mouth of Coal Creek Canyon. Steinheimer virtually "lived" in his various Land Rovers, Travel-Alls, and vans over the years, sometimes car camping for weeks on end during exhausting but exhilarating road trips, usually with one or two companions along. When this image was made, during an inspired three-week solo junket around the intermountain West in 1964, ones senses Steinheimer parked the previous night near the tracks and went to bed with the intention of photographing trains the next morning.

**PLATE 5** Southern Pacific, *Starlight* passenger train, Glendale, California, circa 1949.

Steinheimer's earliest attempts at after-dark photography involved time exposures of the passenger trains that stopped at SP's Glendale Station; night photography became an important part of his life's work.

**PLATE 6** Chicago, Burlington and Quincy northbound freight with SD9 #448 in Sheep Canyon, near Greybull, Wyoming, May 1964.

In an image that duplicates the location of a CB&Q publicity photo Steinheimer glimpsed in Lucius Beebe's book *Highball* in 1945, a northbound freight rolls through Sheep Canyon. This line was constructed by the CB&Q in 1909 to provide a shortcut to the Northwest and tap oil reserves near Casper, Wyoming.

**PLATE 7** Denver and Rio Grande Western 2-10-2 #1403 with coal drag, Thistle, Utah, December, 1951.

At Spanish Fork Canyon just outside Thistle, Utah, Steinheimer makes one of his early masterpieces of the steam era. The temperature, hovering at –28 degrees, has turned bitterly cold despite the bright sunshine. A D&RGW drag freight comprised of empty coal cars and gondolas heads for Soldier Summit; a half mile back a 2-8-8-2 helper shoves the rear end to overcome stiff gradients. This image, a companion piece to the less frequently reproduced one (see plate 145) was the first sheet of 4 x 5 film Steinheimer exposed that day.

**PLATE 8** Southern Pacific's *Starlight* passenger train, Glendale, California, circa 1951.

In an early example of off-camera synchronized flash photography for Steinheimer that predates O. Winston Link's work by four years, the SP's *Starlight,* a secondary passenger train on a eleven-hour leisurely overnight schedule to San Francisco, gets a roll out of Glendale, California, amid the drama of stack talk and panting exhaust, all nobly captured in the flashbulb's glare on Super XX film.

**PLATE 9** Union Pacific, 4-6-6-4 Challenger pilots a freight at sunset, near Rawlins, Wyoming, 1953.

A UP 4-6-6-4 Challenger races across the plauteaus of Wyoming near Rawlins, engine and smoke plume silhouetted against the impending sunset. The Challenger, first introduced in 1936 by the UP, became the favorite mountain locomotive for many of the region's railroads, including the D&RGW, GN, NP, and WP. Many historians have suggested this articulated engine was the high-water mark of steam locomotive technology.

**PLATE 10** Union Pacific's *Park City Local* at Devil's Slide, Weber Canyon, Utah, February 1953.

Steinheimer appreciated the West's naturally occurring geologic forms. Here, while getting a ride on UP's *Park City Local* in February 1953, the photographer steps off the caboose to shoot the Devil's Slide, a landform nineteenth-century western photographers—like Carleton Watkins and William Henry Jackson—recorded on their mammoth wet-plates or in stereograph albumen prints. Many other well-known landmarks, like Castle Butte in Green River, Wyoming, also found their way into Steinheimer's images as he traveled the region. The *Park City Local* was a daily-except-Sunday mixed train that ran out of Ogden behind 2-8-2s.

**PLATE 11** Southern Pacific E9 with *Coast Daylight*, Palo Alto, California, June 1964.

Steinheimer practiced for over a year from the rooftop of his home that abutted the SP main line in Palo Alto, trying to master the tripod-mounted pan shot. Here, in a early attempt he deemed a failure, an E9 painted in scarlet and gray guides the northbound *Coast Daylight* past the camera just a few feet from the photographer's backyard fence. This home, adjacent to the SP main line, would be one of five residences over a twenty-year period Steinheimer would have in close proximity to the tracks.

**PLATE 22** Union Pacific station platform and milk cans, Cascade, Idaho, February 1964.

In 1964 Steinheimer made a brief foray along Union Pacific's Idaho Northern branch line out of Emmet. Here, milk cans rest on the platform at Cascade, Idaho, in a portrait from an earlier era, when local mixed trains stopped at rural depots to exchange goods.

**PLATE 23** Western Pacific westbound freight with F7 #919D, Smoke Creek, Nevada, 1965.

On a rainy May day in Nevada's Smoke Creek Desert a westbound Western Pacific freight train bends to its task beneath a leaden sky. A quartet of EMD F7s, led by the venerable #919D, hustle this time freight toward Oakland, California—the WP's western terminus. Steinheimer, known for traversing "hard country" in various four-wheel drive vehicles over the years, drove to this remote location and made camp for the night with his then-wife, Nona, and friend Ron Turner.

**PLATE 24** Denver and Rio Grande Western symbol freight SFO at Arena, Colorado, April 1964.

A D&RGW freight descends "Big 10" Loop and the Rocky Mountain Front Range at Arena, Colorado. These sweeping, panoramic curves—a marvel of railroad engineering originally designed to enable trains to gain elevation in a short distance—captivated passersby earlier in the twentieth century, making it a prime location for Sunday afternoon picnics and train watching. Hence the name "Arena."

**PLATE 25** Union Pacific freight train with DD40Xs, Echo Canyon, Utah, early 1970s.

A pair of DD40Xs, spliced by a smaller EMD unit, muscles a freight train of TOFC down Echo Canyon toward Ogden, Utah. The DD40Xs, affectionately known as "Big Jakes" or "Centennials," generated 133,766 lbs. tractive effort and cost (in 1969 dollars) over a half-million dollars each to build. Steinheimer, in love with the geography of the West, often made dramatic landscapes a prominent feature in his photographs. Ironically, the DD40Xs—considered the world's largest locomotives—appear lilliputian next to the canyon's bold escarpment, as did the previous generation of Big Boy and Challenger steam engines. The UP, like many of the West's Class-1 railroads, never shied away from exploring powerful locomotion to move freight across such vast distances.

**PLATE 26** Midland Continental freight train, Wimbledon, North Dakota, November 1957.

Steinheimer's eye for graphic composition comes into play at Wimbledon, North Dakota, during November 1957. A smash board, protecting the Soo Line crossing, stands stark against wintry sky, forming a natural frame for the Midland Continental's Alco RS-1 1,000hp road-switcher and its diminutive consist. The MC, incorporated in 1909, was a sixty-eight-mile Dakota short line that specialized in bridge traffic. After years of deferred maintenance and unpredictable wheat crops, the railroad ceased operation in 1970.

**PLATE 27** Great Northern brakeman signals with flare, Gold Bar, Washington, May 1964.

It's raining on the west slope of Stevens Pass in the Cascades of Washington and an eastbound Great Northern freight, suffering a hotbox, stops in the small yard at Gold Bar. A brakeman in quintessential railroad garb signals with a flare after locating the freight car to be "set out."

**PLATE 28** Chicago, Burlington and Quincy depot, Powder River, Wyoming, May 1964.

With the train order board in the all-clear position, sunset falls on rails questing westward; the simple, four-square outlines of CB&Q's Powder River, Wyoming, depot stand quietly by as no trains seem imminent. Driving on Wyoming State Highway 20 the next morning, Steinheimer will head northwest into the Wind River Canyon and commit to film great images that capture its rugged geologic features, rushing rapids, and cliff-hanging train operations.

**PLATE 29** Milwaukee Road Little Joes switch cars at Sappington Hereford Ranch, Willow Creek, Montana, 1964.

Cowboys corral stray cattle at the Sappinton Hereford Ranch in Willow Creek, just west of Three Forks, Montana, while two Milwaukee Road Little Joes set out livestock cars in the distance from eastbound train #264. In the summer of 1805, a Blackfoot Indian woman named Sacajawea helped guide the Lewis and Clark expedition through this region; the CMSt.P&P main line through Montana and Idaho often hove close to this trail blazed almost two centuries ago.

**PLATE 30** Northwestern Pacific *Redwood* passenger train, Scotia, California, May 1957.

The Northwestern Pacific's *Redwood* passenger train #3 hugs the Scotia Bluffs along the Eel River on a somber-toned California North Coast day. The NWP, an SP subsidiary, hauled lumber and forest products from the mills below Eureka; the region's moist climate was directly attributable to the natural growth of redwood in the area. The train is southbound and will see Tiburon and the San Francisco Bay before nightfall. A customized SP "mars" headlight graces the nose of SD9 #5327, a first-generation diesel seen here in "black-widow" paint—a color scheme later supplanted by SP's scarlet and gray.

**PLATE 31** Milwaukee Road freight train on trestle, near Pee Dee, Idaho, April 1972.

A Milwaukee Road freight rolls across the Pee Dee viaduct near Chatcolet Lake in Idaho. This bridge, a fifteen-span open deck trestle, was one of several found on CMSt.P&P's unelectrified "gap" between Avery, Idaho, and Othello, Washington. Today, in 2004, the St. Maries River Railroad runs SW1200s with lumber trains over this twenty-mile remnant of former MR transcontinental main line. The STMA interchanges with the UP at Plummer Junction.

**PLATE 32** Southern Pacific eastbound freight with Fs, Palisade, Nevada, September 1962.

Running wrong-direction, an SP freight with a quartet of F-units growls through Palisade Canyon operating eastbound on the westbound main line. This is joint trackage territory as the WP and SP shared both lines through this region dating from 1924 agreements. The photographer's well-worn Willys station wagon is in the foreground. The former town site of Palisade, the northern terminus of the narrow gauge Eureka Nevada Railway—a line abandoned in 1938—is to the right.

**PLATE 33** Union Pacific "Big Blow" Turbine X-20 and freight train, Cheyenne, Wyoming, 1964.

Union Pacific's "Big Blow" Turbine X-20 pulls a mixed freight into Cheyenne, Wyoming, off the Nebraska Division. The gas turbines were part of the UP's on-going tradition of utilizing the largest, most powerful locomotives to move tonnage. The UP fleet numbered fifty-five at one point; the first one was tested on the railroad in 1949. Built by General Electric, the initial ten turbines, numbered 51-60, arrived on the property in 1952. Number X-20 was part of a final order placed between 1958 and 1961 and was rated at 8,500hp.

**PLATE 34** Chicago, Burlington and Quincy train #29, Wind River Canyon, near Thermopolis, Wyoming, 1964.

CB&Q train #29 with an E9 on the point emerges from a wood-lined tunnel portal in Wyoming's Wind River Canyon north of Greybull. This area, once explored by John C. Fremont in 1842 on a government survey, was also the scene for his initial daguerreotype experiments—which failed due to his lack of experience and the complexities of mastering the process in the field. Steinheimer, utilizing modern film and equipment 122 years later, had no such technical problems.

**PLATE 35** Southern Pacific commuter looking out window of Harriman Coach, Palo Alto, California, 1955.

A bored passenger gazes idly out a Harriman coach window as it rolls by Palo Alto, California, impatient for his daily commute to be over. Steinheimer, living in close proximity to SP's San Francisco commute line, exploited its daily rush hour "parades" as fodder for ceaseless photographic experimentation and narrative possibilities. The commutes provided "backdoor" access to the world of railroading.

**PLATE 36** Milwaukee Road, welder and boxcabs, Cedar Falls, Washington, 1967.

A welder walks away from GE boxcab #E30, adding a forlorn feeling to an already overcast day in Cedar Falls, Washington. The fifty-year-old "pelicans" required occasional repairs in order to keep them operational; one senses their longevity was never a burden to the mechanical department. This set, used on the Milwaukee Road's Coast Division, provided additional helper power for trains running over the Cascades between Othello and Tacoma. Electrification on this segment of CMSt.P&P ended in 1972.

**PLATE 37** Colorado and Southern #900 on turntable, Cheyenne, Wyoming, February 1956.

Colorado and Southern #900, a bulky 2-10-2, rolls off the turntable at Cheyenne, Wyoming, in this wintry view. The C&S was incorporated in 1898, eventually operating main lines that extended from Orin Junction, Wyoming, to Fort Worth, Texas. The CB&Q purchased a two-thirds controlling interest in December 1908. The C&S became a part of the BN in 1981.

**PLATE 38** Southern Pacific freight at Cantera Loop near Dunsmuir, California, February 1964.

After a three-mile hike in snowshoes through chest-high drifts with his wife and children, Steinheimer records a Southern Pacific westbound symbol freight descending Cantera Loop above Dunsmuir, California, in the winter of 1964. Steinheimer, characteristically immune to long treks in inclement weather, made pictures taken under such adverse conditions synonymous with his name; for him, hardship and success were twinned concepts.

**PLATE 39** Great Northern *Empire Builder* departs King Street Station, Seattle, Washington, May 1964.

In a photograph more reminiscent of Philip Hastings than Richard Steinheimer (Hastings liked to thrust a compositional element into the foreground of his images), Dick makes a shot of the *Empire Builder*, framed by ornate grillwork, departing Seattle's King Street Station for Chicago. King Street Station opened in 1906 and featured a 240-foot campanile whose chiming bell could be heard throughout the business district. Northern Pacific passenger trains shared the facility with the Great Northern at the time of this photograph.

**PLATE 40** Southern Pacific freight on Donner Pass, near Cisco, California, 1972.

Southern Pacific SD40 #8439 with a westbound symbol freight descends Donner Pass below Cisco, California. Steinheimer routinely hiked and snowshoed all over Donner Pass, becoming as familiar with this mountainous region as Ansel Adams was with Yosemite; he perhaps knew it better that any other person. For this image, of a train crossing Butte Canon Bridge, Steinheimer employed high-contrast printing paper to enhance the graphic qualities of locomotive, girder work, and Sierra Nevada terrain.

**PLATE 41** Northern Pacific FT #5402A with freight train, Pasco, Washington, June 1965.

Stunning symmetries of signal, track, and approaching train form an arresting composition as Northern Pacific FT #5402A, with forty-seven cars in tow, enters the yard at Pasco, Washington. This Northern Pacific FT engine was one of a diminishing example of the first practical freight diesel locomotives delivered by EMD in the early 1950s. The NP ran the largest fleet of FTs at the time of this photograph, but within a few years these locomotives became trade-in candidates for higher horsepower units: change and transition being the constants in the technology-driven world of railroading.

**PLATE 42** Southern Pacific cab-forwards at Taylor Yard Roundhouse, Los Angeles, California, circa 1949.

This early Steinheimer photo captures the men and machines, a preoccupation that would inform his work over the next fifty years. The scene is Taylor Yard Roundhouse in Los Angeles in the late 1940s. The brute, massive architecture of cab-forwards provides a dramatic backdrop to the pipe fitters, laborers, and mechanics moving among the colossal 4-8-8-2 articulated locomotives.

**PLATE 43** Southern Pacific steam engines being scrapped with 2-10-2 in background, Richmond, California, November 1956.

The scrap yard of Richmond, California's Southwest Welding & Manufacturing Company becomes the final destination for many of the SP's Bay Area steam locomotives in the 1950s. In a scene evoking Dante's *Inferno*, the scraper's torch is hard at work deconstructing and dismantling a string of 2-10-2s; the work takes on an inverse assembly-line approach as larger pieces are cut continually into smaller ones.

**PLATE 44** Leased Milwaukee Road 44-ton #1701 outside White Sulpher Springs and Yellowstone Park engine house, Ringling, Montana, April 1956.

Newly married, Steinheimer and wife Nona took a three-week trip to Montana in April 1956—the second journey for the photographer along the Milwaukee Road. Here, a leased CMSt.P&P center cab gets ready for the mornings activities on the twenty-three-mile WSS&YP—a short line incorporated in 1910. Stein had a deep affection for the quainter railroad operations in the West, eagerly seeking them out throughout his career.

**PLATE 45** Longview, Portland and Northern freight train parked at crossing, Battleground, Washington, April 1963.

With the crew temporarily "gone to beans," a Longview, Portland and Northern train with Alco #112 pauses at Battleground, Washington, during a nocturnal run on the thirty-three-mile ex-NP Yacolt Branch to Chelatchie. The LP&N, incorporated in 1922, is a wholly owned subsidiary of International Paper Company. As of 2004, it operates 3.6 miles of track and transports yearly 3,000 carloads of lumber and forest products.

**PLATE 46** Southern Pacific, lone worker and roundhouse, San Luis Obispo, California, mid-1950s.

A roundhouse worker crosses the temporarily vacant tracks of Southern Pacific's fifteen-stall San Luis Obispo roundhouse in the mid-1950s. Erected in 1894, the roundhouse eventually become a busy mid-way division point with the Coast Route's completion in 1901, supplying a multitude of helper engines and crews to lift *Daylights* and hotshot freights over Cuesta Grade and the Santa Lucia Range.

**PLATE 47** Northern Paciifc 2-6-0 #1110 switching cars, Chehalis, Washington, April 1956.

In a bucolic small town scene, NP #1110, a little 0-6-0 locomotive, performs switching chores in Chehalis, Washington. A Pontiac waits at the crossing for the switchman to give the all-clear. A light plume of smoke suggests the engineer, who's riding a cab armrest, has just released the Johnson Bar, backing the diminutive steamer into a cut of boxcars.

**PLATE 48** Southern Pacific freight yard in snow, Dunsmuir, California, February 1959.

Looking over a fur tree and down on a boxcar-filled, snow-covered Dunsmuir yard, Steinheimer captures an alluring winter railroad scene. Out taking pictures when most photographers wouldn't, "Stein" took advantage of blizzard forecasts all over California and the intermountain West to make memorable climatological images that became his trademark.

**PLATE 49** Colorado and Southern train #8-28 with E5 #9955, Pueblo, Colorado, 1964.

Colorado and Southern passenger train #8-28 with E5 #9955 rolls beneath a highway overpass as it enters the station yard at Pueblo, Colorado. An elderly gentleman pays homage to the time-honored tradition of watching trains go by. The Pueblo Union Depot served Santa Fe, Missouri Pacific, Denver and Rio Grande, Rock Island, and Colorado & Southern. This massive structure was erected in 1889–90 and was designed by Sprague & Newall of Chicago, Illinois.

**PLATE 50** Stockton, Terminal and Eastern engine house, Stockton, California, February 1970.

Tule fog blankets the Central Valley and the Stockton, Terminal & Eastern's engine house as a crewman heads to work. The ST&E was established in 1908 and still operates twenty-five miles of track in 2004, connecting with the UP, BNSF, and Central California Traction.

**PLATE 51** Atchison, Topeka and Santa Fe westbound *El Capitan* at dusk, Yampai, Arizona, 1968.

The Santa Fe *El Capitan,* resplendent with high-level full-dome cars, rolls into twilight, traveling at 68 mph on the high iron near Yampai, Arizona, in 1968. Santa Fe often combined the *Super Chief* and *El Cap.*

**PLATE 52** Western Pacific's *California Zephyr* at Fremont, California, March 1970.

The westbound *California Zephyr* makes its final curtain call at Fremont, California, on March 17, 1970. Silhouetted against the gathering twilight, a young boy intuits an era is passing; a train jointly operated by the CB&Q, D&RGW, and WP for twenty-one years is making its last run. Steinheimer, at the end bearing witness to an evolving world of railroading, tallied more losses than gains. On the negative's envelope at the DeGolyer Library, the photographer scribbled the words "SAD DAY."

**PLATE 53** Two Southern Pacific switchmen looking over lists for "Zipper," Los Angeles, California, 1952.

At SP's Alameda Street Yard in Los Angeles, a pair of railroaders check their switch list and waybills prior to the departure of the Coast Merchandise West (also known as the "Zipper"), Southern Pacific's hotshot 13 1/2-hour overnight "less-than-carload" freight train to the Bay Area. Southern Pacific aggressively marketed this expedited service in fierce competition with trucking lines: it truly was the FedEx of the 1950s.

**PLATE 54** Colorado and Southern 2-10-2 #900 in roundhouse, Cheyenne, Wyoming, October 1957.

In an intimate locomotive portrait, Colorado and Southern 2-10-2 #900 gets ready to leave the roundhouse in Cheyenne for a freight run north to Horse Creek and Orin Junction,Wyoming.

**PLATE 55** Oregon and Northwestern AS616 locomotive and log train, near Burns, Oregon, 1960s.

A Baldwin AS616 locomotive of the Oregon and Northwestern Railroad moves a log train through fresh fallen snow near Burns, Oregon. The O&NW was a fifty-mile short line that started life as the Malhuer Railroad in 1923 and interchanged with the UP at Burns, Oregon; the line was abandoned in 1985.

**PLATE 56** San Francisco State Belt Railway Alco switcher being repainted, San Francisco, California, 1958.

A mixture of ladders and scaffolding add sculptural dimension to this photograph of San Francisco State Belt Railway 1000hp Alco S3 diesel receiving a new coat of paint in 1958. The SFSB Ry., a terminal switching line serving the SP, AT&SF, NWP, and WP, operated ten miles of main line and forty-three miles of track along San Francisco's famed Embarcadero waterfront.

**PLATE 57** Southern Pacific freight, as seen from State Highway 99, Cottage Grove, Oregon, March 1967.

An Oregon rainstorm pelts the windshield of Steinheimer's car as he carefully times exposures of a northbound SP freight between swipes of the wiper blades. Steinheimer never hesitated to utilize or incorporate the automobile into his photography. Perhaps, in this case, it was necessary—a convenient way to stay dry while battling stormy weather in the perpetually damp Northwest.

**PLATE 58** San Diego & Arizona Eastern yard scene, San Diego, California, 1952.

Baldwin AS616 road-switcher #5247, wearing the "black widow" paint scheme that adorned many Southern Pacific first-generation diesels, shunts freight cars around subsidiary San Diego & Arizona Eastern's pocket yard near downtown San Diego, California. Outside-braced boxcars and fallen-flag logo bearing forty-footers attest to this photo's post-WWII vintage. Steinheimer captured this excellent tableau of the industrial sublime—an early example of the photographer incorporating the larger environment into his railroad photography.

**PLATE 59** Northern Pacific freight #603 in yard, Jamestown, North Dakota, November 1957.

Northern Pacific westbound train #603 pauses seventeen minutes in Jamestown Yard for inspection, pickup, and crew change. A swirling blizzard, sweeping down from Canada across the Dakotas and Minnesota, has slowed the progress of this Chicago–to–Puget Sound hotshot: the NP's fastest time-carded freight train to the West Coast. After four cars of hogs are tacked on the head end, and the train recoupled, Mandan—107 miles to the west—is the next stop.

**PLATE 60** Denver and Rio Grande Western freight train, west of Antonito, Colorado, 1950s.

In a brilliant backlit exposure, Steinheimer captures the inherent drama of steam railroading as a double-engined Denver and Rio Grande Western narrow gauge freight runs through a wintry landscape west of Antonito, Colorado. Steinheimer typically exposed one-sheet or frame of film for many of his early masterpieces.

**PLATE 61** Mount Hood Railroad, pacing shot of Alco switcher, Dee, Oregon, March 1963.

The Hood River Railroad, which operated south from the UP main line at Hood River, Oregon, was a twenty-one-mile short line incorporated in 1905, independently run and then chartered by the UP in 1968; it ceased freight operations in 1987 and now runs as a popular tourist line. Here Steinheimer captures on 4 x 5 film the daily freight with Alco switcher #50 at speed near Dee, Oregon, during a trip he and Don Sims made to the Northwest. The photograph, a pacing shot, was made from the backseat of a moving vehicle.

**PLATE 62** Northern Pacific FT #5402A with freight train enters yard, Pasco, Washington, 1965.

In a dramatic telephoto portrait of diesel action on the upland plains flanking the Columbia River, Northern Pacific FTs drop down off a steep grade into NP's Pasco Yard. The Buick Roadmaster cab styling of the FTs made them a favorite with crews and railfans. The arrival of these first-generation diesels on America's railroads in 1939 spelled doom for steam locomotion. The NP owned a fleet of ninety-five FTs, second only in number to the Santa Fe. Many eventually became trade-in candidates on GP9s and other second-generaton hood units.

**PLATE 63** Chicago, Burlington and Quincy daily freight, Parkman, Wyoming, 1953.

In a brilliant wintry display of snow and steam, a CB&Q daily southbound train stumbles up the gradients of Parkman Hill, cresting the main line near the Wyoming-Montana border. The lumbering 2-10-2 gets an assist from 4-8-2 #7702. Sheridan, Wyoming, the division point, is a few miles away.

**PLATE 64** Great Northern F7 #278A coupling to train, Seattle, Washington, May 1964.

Utilizing a vantage point atop a railroad embankment, Steinheimer captures a quintessential train operation in all its subtlety: the act of coupling up. Great Northern F7 #278A gently backs onto its train in Seattle, Washington, as a brakeman signals to the engineer, guiding this time-honored procedure.

**PLATE 65** San Diego and Arizona Eastern freight #451 on Goat Canyon trestle, Carriso Gorge, California, 1953.

A Baldwin AS616 road-switcher crawls across the 635-foot-long, 185-foot-high Goat Canyon Trestle with westbound freight #451 in tow. The trestle, one of the largest in the world, was part of a track realignment in 1932 after tunnel 7 burned. The train runs through the deep declivities of the San Diego and Arizona Eastern's Carriso Gorge, some of the most rugged geology traversed by any railroad in North America. The line, originally constructed by sugar magnate John D. Spreckels, and later sold to the SP in 1933, took thirteen years to build. Steinheimer made this shot during a 1953 weekend outing while stationed in the U.S. Navy at San Diego. SP abandoned the route in 1975 after Hurricane Kathleen destroyed several tunnels on the line.

**PLATE 66** Wheel and journal box of a PFE reefer, Roseville, California, May 1965.

Steinheimer captures the graphic simplicity of a Pacific Fruit Express "reefer" wheel and journal box at Southern Pacific's Roseville, California, yard. The journal is open for a "car knocker" to lubricate prior to departure on a Portland-bound Shasta Route freight. Pacific Fruit Express, begun with a fleet of 6,600 cars, was organized by the UP and SP in 1906 to transport perishables to eastern markets. Roseville, in the early 1950s, had the world's largest ice manufacturing plant to service these refrigerated cars.

**PLATE 67** Midland Continental track laborers at grade crossing, Jamestown, North Dakota, 1957.

Under a monotone Dakota sky, three track workers on the Midland Continental dig out a rural grade crossing near Jamestown, North Dakota, in a tableau capturing the honesty of manual labor. Throughout its history, the MC ran on shoestring budgets—continually plagued by revenue shortfalls and deferred maintenance—which often meant running on a less than satisfactory roadbed. The last trains operating in 1968 ran on track with almost no ballast. The line was abandoned in 1970.

**PLATE 68** Bridge and Building foreman inspects track #1 on Southern Pacific main line, east of Donner Summit, California, 1965.

A bridge-and-building foreman traverses snow-covered rails near Southern Pacific's Donner Summit—the highest point on the Overland Route. Inspecting trackage for potential snow or rock slides is a 24-7 process during the winter months in the High Sierra, where annual snow accumulation can easily exceed 400 inches. This view looks down on track #1, near tunnel 7; Donner Lake is seen in the distance. This photo was made while Steinheimer was on a trip with friend David Straussman.

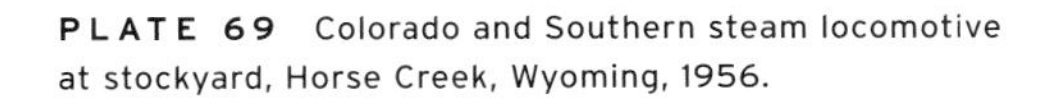

**PLATE 69** Colorado and Southern steam locomotive at stockyard, Horse Creek, Wyoming, 1956.

Transporting animals—whether hogs, cattle, or sheep—was still an ongoing business for the Colorado and Southern in 1956. Here, a herd of pensive cattle wait to be loaded onto livestock cars while, in the background, C&S 2-10-2 #900 switches the small yard at Horse Creek, Wyoming.

**PLATE 70** Southern Pacific's *City of San Francisco*, Applegate, California, 1961–62.

In the right place at the right time: Steinheimer, in a "decisive moment," captures the westbound *City of San Francisco*—"just off the advertised" yet exploding into the picture frame—as it ducks under an eastbound freight near tunnel 26 on the flyover at Applegate, California. Luck, created through relentless hard work and tenacity, would find Steinheimer in countless circumstances "making things happen" over his five decades of photography.

**PLATE 71** Milwaukee Road freight on branch line near Manhattan, Montana, 1964.

Rolling across the "big empty," the lyrical name bestowed by many a Montana writer upon this landscape, Milwaukee Road SW1200 #623 and its four-car train of forty-foot boxcars, heads toward the Gallatin Gateway and Bozeman, Montana, in May 1964. In the foreground lies the NP main line that will soon rumble to the passage of an eastbound *North Coast Limited.* That Steinheimer had a fondness for branch lines, short lines, and lesser-known railroad operations is evident by this beautifully composed paean to everyday insignificance elevated to art.

**PLATE 72** Chicago, Burlington and Quincy freight train #75 in Wind River Canyon, near Thermopolis, Wyoming, April 1964.

The wild Wind River Canyon, deep within the Bridger Mountains of north-central Wyoming, plays host to the passage of train #75, a CB&Q daily mixed freight bound for Laurel, Montana. Five SD9s power a train dwarfed by the terrain that inspired explorer John C. Fremont, who attempted daguerreotypes of this imposing landscape in 1842. Fittingly, in honor of his photographic predecessors, Steinheimer made several memorable images here.

**PLATE 73** Union Pacific's *Los Angeles Limited* at Sullivan's Curve, Cajon Pass, California, 1952.

In an early photograph where Steinheimer made the natural environment a key player in the composition, Union Pacific's *Los Angeles Limited* rounds Sullivan's Curve on Cajon Pass in 1952. A 5000-class 2-10-2 helper assists a trio of Alco FAs as they lift the train through the rugged geology of this famed California railfan landmark.

**PLATE 74** Dan Murray, Milwaukee Road conductor, Malden, Washington, May 1964.

With fedora in place, pipe and overnight satchel in hand, Dan Murray, a Milwaukee Road conductor at Malden, Washington, waits for incoming train #262. This is possibly the photographer's finest portrait. Steinheimer openly sought the camaraderie and friendship of railroaders he encountered; he was accustomed to making some type of picture of almost every worker he came in contact with—and routinely gave prints away as an act of gratitude. It was not uncommon to see his photographs on depot office walls across the West.

**PLATE 75** Southern Pacific locomotive shops, Sacramento, California, 1956.

As the Central Pacific grew, its isolation from eastern centers of locomotive manufacturing dictated that it erect its own shops and build many of its own locomotives. CP's first engine was the diamond-stacked eight-wheeler #173, the last a 0-8-0 switcher. In the intervening years, over 200 steam engines were erected on the sprawling Sacramento shops' 200-acre facility. In this view, a 2-10-2 remains alive in the twilight of SP steam, one of the more fortunate members of a locomotive class that once numbered nearly 200. Today, in 2004, part of the shop complex is owned by the CSRM.

**PLATE 76** Missouri Pacific's southbound *Colorado Eagle*, Palmer Lake, Colorado, April 1964.

The operator stands beside the Santa Fe depot at Palmer Lake, Colorado, with its quaint gingerbread architectural elements, and does a traditional roll by inspection of a southbound Missouri Pacific *Colorado Eagle*. A lone EMD E-unit pilots this train on its journey toward Kansas City. The MP had trackage rights over the D&RGW/Santa Fe/C&S joint line between Denver and Pueblo.

**PLATE 77** Southern Pacific F7 # 6229 with time freight, Antelope, California, early 1960s.

SP F7 #6229 with a time freight rolls past budding apple trees north of Roseville, California, in the early 1960s. Steinheimer utilized time exposure and remote flash to create this image. The artificial light, placed strategically behind the tree and hidden from view, simultaneously lights the diesel units and throws a beautiful rim-lit luminosity on branches and leaves. This is one of several frames he exposed that night: for another, see Beebe's *Central Pacific and Southern Pacific Railroads,* page 610, where additional lighting was added. Steinheimer became expert at visualizing night photographs.

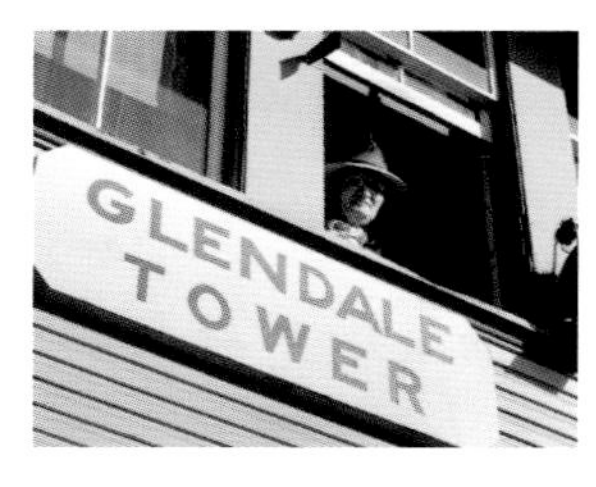


**PLATE 78** Southern Pacific's Glendale Tower, Glendale, California, 1950.

Glendale Tower was a place of refuge for the young Steinheimer—frequent visits there to work the twenty interlocking levers or two derails (which the railroaders would allow him to do)—brightened his childhood. Mr. Peterson, a towerman who befriended the photographer, poses for his portrait framed in an upstairs window.

**PLATE 79** Western Pacific symbol SWG freight with Fs, Stockton, California, March 1965.

Western Pacific's SWG (Santa Fe, Western Pacific, and Great Northern) finally gets permission to cross the diamond and Santa Fe tracks. The AT&SF *Super Chief* has just cleared the interlocking and the SWG begins its journey through the upper San Joaquin Valley and north toward the Feather River Canyon. The three railroads utilized the "Inside Gateway," a late-line extension completed in 1931 to compete with the SP's Cascade Route.

**PLATE 80** Great Northern westbound freight train with F7 #460A, near Browning, Montana, May 1964.

Running across the high plains of the Blackfoot Indian reservation near Browning, Montana, a solid A-B-B-A set of F7-units leads a Seattle-bound consist toward a wooden snowshed. Erected to fight winter snow accumulation in cuts or depressions, these snowsheds were an unusual architectural feature along GN's main line east of the Rockies. Due to fire hazard and maintenance concerns they were dismantled in the late 1960s.

**PLATE 81** Southern Pacific freight train at Crystal Lake snowsheds, near Cisco, California, 1967.

An Alco Century C-628, built in 1965, pilots a train downgrade and westbound past the Crystal Lake snowsheds in 1967. Sheds, constructed on Donner Pass to combat snow accumulation, date from the earliest days of the Central Pacific; there were almost forty miles of them at one time. When this shot was made, the SP had a summer program of converting wooden sheds to reinforced concrete—a move instigated after the disastrous Norden fire in November 1961 destroyed over 4,300 feet of shed, track, and ties. The snowsheds would hold a special fascination for Steinheimer; in fact, the Norden operator's office is where I first encountered his prints decorating the walls in a railroad workplace. The trainmen loved his work and proudly displayed it; Stein was on a first-name basis with many SP employees on "The Hill."

**PLATE 82** Dutch Brantz, San Diego & Arizona Eastern roundhouse foreman, Calexico, California, 1953.

Dutch Brantz, the cigar-smoking, bilingual roundhouse foreman at Calexico, chalk in hand, juggles roster requirements to keep trains moving over SP's SD&AE and Inter-California Railway subsidiaries. The following year Imperial Valley's low desert rails would succumb to dieselization. Steinheimer became familiar with the rail operations in this remote part of southeast California while serving in the U.S. Navy at San Diego during the early 50s.

**PLATE 83** Southern Pacific 0-6-0 switch engine #217, Oakland, California, 1956.

The Southern Pacific had a knack for building home-grown switch engines. This example, found near the Oakland roundhouse in 1956, seems like a piece of machinery from a H. G. Wells novel. Engine #217, built by Brooks around the turn of the century, was originally 0-6-0 #1295, one of twenty S-8 class engines. Converted to a saddle tank engine by the SP in 1940, it is here shunting a much larger articulated, cab-forward locomotive just out of the frame.

**PLATE 84** Carbon County Railway SW1200s and empties, near Columbia, Utah, 1960s.

In a composition verging on abstraction, a pair of Carbon County Railway SW1200s lead a string of empties—the first of two daily trains—toward the mine at Columbia, Utah, in the 1960s. This eleven-mile short line, constructed in 1923, connected to D&RGW's Sunnyside branch and was abandoned in 1982. A slag heap smolders in the distance, lending an interesting, other worldly tone to the overall image.

**PLATE 85** Tucson, Cornelia and Gila Bend Railway freight train, Black Gap, Arizona, 1952.

Seen here at Black Gap, Arizona, a Tucson, Cornelia and Gila Bend EMD switcher scoots past a saguaro cactus on its daily run to Ajo, and the copper mines and smelters of the Phelps Dodge Corporation. The line, which suspended operation in 1984, was a forty-three-mile short line that ran through the barren Sauceda Mountains.

**PLATE 86** Union Pacific brakeman on top of cattle car, Cajon Pass, California, early 1950s.

Bound for the meatpacking plants of Los Angeles, the Union Pacific's "Daily Livestock Special" from the early 1950s stops atop Cajon Pass for the brakeman to set retainers for the descent into San Bernardino and the L.A. Basin. The amount of sheep and cattle moving by rail declined in the mid-1950s as road improvements made shipping by truck a feasible alternative.

**PLATE 87** Denver and Rio Grande Western depot and train, Spanish Fork, Utah, 1964.

A D&RGW ore train piloted by Geep #5912 passes the Spanish Fork depot on the Tintic branch south of Provo, Utah; Steinheimer loved obscure, remote branch lines.

**PLATE 88** Southern Pacific & Western Pacific crossing, Flanigan, Nevada, June 1961.

Long before it became popular to take railroad pictures without trains, Steinheimer captured the essential railroad environment absent, ironically, of any motive power. Here, in Nevada's Smoke Creek Desert, two main lines—belonging to the WP and SP—intersect momentarily. Toward the southeast, we see SP's Fernley–to–Pyramid Lake line, abandoned in 1963 after a renegotiated trackage-rights agreement with the WP made it moribund.

**PLATE 89** CB&Q northbound freight with SD9 #448 in Sheep Canyon, near Greybull, Wyoming, 1964.

In an image that practically duplicates the location of a CB&Q publicity photo Steinheimer first glimpsed in Lucius Beebe's book *Highball* in1945, a northbound freight piloted by SD9 #448 rolls through Wyoming's resplendent Sheep Canyon near Greybull on the way to Billings, Montana. The majestic qualities of the inter-mountain West would become an integral backdrop for Steinheimer's photography. In fact, one assumes he drew direct early inspiration from this specific image. It was one of two in *Highball* that showed the train in a bold geology; the other, a Beebe photo on the C&S Climax branch, was also an image Steinheimer paid homage to (see *Highball*, page 203, and *Backwoods Railroads of the West*, page 109).

**PLATE 90** Southern San Luis Valley Railroad caboose at sunset, Blanca, Colorado, January 1961.

This weather-worn caboose resting in solitude at Blanca, Colorado, captures the despair of a moribund short line, while the Sangre de Christo Mountains hover as sentinels in the distance. The Southern San Luis Valley (originally San Luis Valley Southern) once served agricultural interests in the San Luis Valley and ran 2-8-0 Consolidations to Jaroso that connected with Rio Grande's *New Mexico and Colorado Express.*

**PLATE 91** Southern Pacific's *Del Monte* with GP9 #5600, San Francisco, California, December 1964.

San Francisco's downtown skyline looms as Southern Pacific's southbound *Del Monte* gets a roll out of Third and Townsend, seen here on the first big curve along Seventh Street south of the station. In this view, Steinheimer utilizes a unusual piece of equipment—his "super-tele"—to compress the film plane, creating drama and tension. Steinheimer typically gave himself assignments such as these to push his notions of what was possible. He photographed the commute parade in the same location for several hours; his use of a 300mm lens in the mid-1960s was indeed novel and experimental.

**PLATE 92** Atchison, Topeka and Santa Fe Engine Terminal, Chicago, Illinois, November 1965.

A wide variety of differing locomotive types surround the turntable at the south end of Santa Fe's Chicago yard as Alco, EMD, and F-M power dot the scene. Steinheimer made frequent trips to the Windy City for his employer, Fairchild Semiconductor, in the 1960s, making quick forays around the area when free time allowed.

**PLATE 93** Southern Pacific 4-6-2 #2484 and freight train, Oakland, California, 1956.

SP streamlined Pacific #2484 heads out of Oakland, California, bound for Tracy with a drag freight. Steinheimer routinely relied on available light to create certain moods; a portion of his nightwork was made without flash. He was fond of saying: "f16 and a five minute time exposure on Tri-X will accomplish anything you want to do."

**PLATE 94** Union Pacific freight train with DD35 #71, Cima, California, 1977.

Union Pacific DD35A #71 waits in the siding to meet an opposing train near Cima Hill in California's Mojave Desert. A thunderstorm—not an infrequent occurrence in high desert country—gathers just beyond the locomotive; a full moon washes over the desert landscape. Steinheimer worked between the years 1975 and 1978 as a writer and audiovisual specialist for Ford Aerospace at Twentynine Palms, California—a locale that gave him ample opportunity to familiarize himself with UP and AT&SF desert operations.

**PLATE 95** Denver and Rio Grande freight train near Provo, Utah, 1964.

A D&RGW sixteen-car ore cars rolls beneath the almost apocalyptic wintry grandeur of Utah's Wasatch Range south of Provo. The Tintic branch, built standard gauge in 1891–92 to tap the rich mineral deposits around Silver City, was initially forty miles long and operated from Springville to Silver City. The line, inherited by the UP after it merged with SP/D&RGW interests in 1996, was sold in 2002 to the Utah Transit Authority for light rail commuter use to ease suburban congestion along the I-15 corridor.

**PLATE 96** Southern Pacific troop train with GS-4, near Casmalia, California, 1955.

A farmer plows a field in preparation for next spring's alfalfa crop, as a Southern Pacific "Daylight" GS-4 lays whispy smoke patterns back over a troop train—perhaps bound for Camp Roberts—near Casmalia, California. The "GS" designation for these 4-8-4s originally meant "Golden State," but was later changed to "general service."

**PLATE 97** Atchison, Topeka and Santa Fe freight near Monolith, California, early 1970s.

A Santa Fe westbound freight grinds uphill toward Monolith after a long pull on a 2.2 percent grade out of Mojave, California. Steinheimer takes advantage of a long lens and uses telephone poles—a distracting element to most railroad photographers—to create an arresting composition. As he once said: "I always wanted to take my photography a little further."

**PLATE 98** Denver and Rio Grande freight exiting Moffat Tunnel, East Portal, Colorado, April 1964.

New D&RGW GP30s thunder from the east portal of Moffat Tunnel as spring snow flurries fly off the Rockies. The train consist stretches partially back through Moffat's six miles of darkness. This tunnel beneath the Continental Divide was completed in 1927. When Steinheimer made this image, the Rio Grande was a feisty competitor of the Union Pacific to the north and Santa Fe to the south. D&RGW's main lines through the Rockies doomed it to slower delivery times and more costly maintenance than its larger, flatland rivals. As of 2004, the former Moffat Road is a secondary main line owned by the UP that sees frequent coal traffic.

**PLATE 99** Southern Pacific's *City of San Francisco* arrives at Oakland, California, November 1962.

In an unusual composition, Steinheimer catches the arrival of the *City of San Francisco* at Oakland's 16th Street Station. This picture was taken for the United States Information Agency; the assignment called for "shots of people getting on a passenger train somewhere." While Steinheimer claimed the agency was "less than happy with the results," the photographer liked the pictures well enough to retain them in his files.

**PLATE 100** Southern Pacific, hostler and Alco #9018 at diesel shops, Roseville, California, 1966.

The chiseled face of a hostler leans out the cab window of #9018—one of three mammoth Alco DH-643 diesel hydraulics owned by the SP—while performing a move in the shops at Roseville, California, in 1966. Despite its impressive size at seventy-five feet long and weighing 400,000 lbs., this locomotive was eclipsed by the heavier DD35 and U50 members of SP's diesel fleet. The hydraulic era ended on the SP in 1973 when these Alcos were deleted from the roster.

**PLATE 101** Northern Pacific fireman taking on water for 2-8-2 #1708, Auburn, Washington, April 1956.

A fireman fills the tender of NP #1708, a 2-8-2 type in Auburn, Washington, in April 1956. His relaxed pose adds to the picture's graphic qualities. Steinheimer, never overly invested in "form for form's sake," nonetheless had a good eye for recognizing strong, straightforward, and well-balanced compositions.

**PLATE 102** Midland Continental freight train, Jamestown, North Dakota, November 1957.

A Midland Continental mixed train piloted by an RS-1 pulls into the outskirts of Jamestown, North Dakota, to switch cars before going on to the depot. The photographer stood atop a boxcar beneath a grain elevator to shoot this minimal prairie landscape. Steinheimer often scaled railroad equipment to gain height or placed his camera at ground level to explore different perspectives.

**PLATE 103** Southern Pacific eastbound freight with 2-10-2 #3652 leaves Colton Yard, Colton, California, 1949.

In an early night photograph, Steinheimer captures a local freight about to depart Colton Yard. On the point is #3652, a 2-10-2 with a Vanderbilt tender. Colton, a busy point on the railroad since its inception in July 1875, remains so today for freights operating over the Palmdale cutoff, built in 1967 to bypass the congested Los Angeles Basin.

**PLATE 104** Union Pacific freight train with F7 #1433, Horseshoe Bend, Idaho, February 1964.

Union Pacific F7 #1433 crosses a tributary of the Payette River north of Horseshoe Bend, Idaho, during the winter of 1964. The train is running on Union Pacific's Idaho Northern Branch—once part of UP's Oregon Short Line.

**PLATE 105** Man and wigwag signal at crossing, Lodi, California, February 1971.

A pedestrian crosses the SP tracks beneath a fog-enshrouded wigwag signal. Steinheimer often made photographs in unusual weather—Central Valley tule fog was a favorite—a climatological condition most evident during the winter months.

**PLATE 106** Union Pacific's *City of Los Angeles*, Green River, Wyoming, July 1952.

Union Pacific's *City of Los Angeles* pauses momentarily for a station stop at Green River, Wyoming, allowing passengers to disembark for a stretch and a smoke. While UP's transcontinental rails reached Green River on October 1, 1868, the town would not become an important division point (with roundhouse, machine shop, and depot) until 1872.

**PLATE 107** Southern Pacific, steam and diesel freight action, Port Costa, California, mid-1950s.

A semi-aerial view of dramatic Bay Area railroading action is captured by Steinheimer at Port Costa, California. Cab-Forward #4213 heads up a Roseville-Oakland manifest, as Alco S2, painted in SP's black and orange tiger stripes, switches cars to local refinery tank farms. This industrial tableau signaled "Stein's" growing awareness of the steam/diesel transition taking place within the railroad industry and exemplifies his willingness to document it, despite being a railfan "scarred by the 50s."

**PLATE 108** Southern Pacific Harriman coaches, San Francisco, California, January 1965.

Steinhiemer catches the tail ends of two Harriman suburban coaches sputtering steam on the platforms of SP's Third and Townsend depot while awaiting the daily late-afternoon onslaught of humanity. SP gave up the commute business in 1980; CalTRANS now operates the service between San Francisco and San Jose.

**PLATE 109** Union Pacific Gas Turbine #54 and freight, Wahsatch, Utah, February 1953.

Sunset unfolds to the west as the operator at Wahsatch, Utah, bravely stands trackside to hand orders up to a passing UP freight with gas turbine #54 on the point. The "big blows," so named because their engines roared like jets, ran mostly between Council Bluffs, Iowa, and Ogden, Utah. The gas turbine experiment, instigated by the UP in 1948 in its unending quest for larger, more efficient motive power, lasted until 1970.

**PLATE 110** Great Northern's *Empire Builder*, East Glacier, Montana, May 1964.

Parked beside Montana's Highway 2—known as the "high line"—Steinheimer grabs a shot of the westbound *Empire Builder* crossing Two Medicine Bridge near East Glacier. The bridge, at 210 feet, was the highest on the GN's transcontinental main line. Ahead is the climb to Marias Pass and splendid views of Glacier National Park.

**PLATE 111** Union Pacific freight train with reefers, Ogden, Utah, October 1952.

A UP freight train consisting of a solid "reefer" block of fruit bound for eastern markets rolls past the flagman through Ogden's industrial backyard. F7A #1479 had been on the property less than a year. Grain elevators and kerosene switch lanterns frame this scene; on the rear end is a thundering 4-8-4.

**PLATE 112** Northern Pacific laborer and RS-1 locomotive, Duluth, Minnesota, October 1957.

A railroad laborer uses the locomotive's air brake system to pump sand into the sandbox for traction in a photograph made on the Northern Pacific in Duluth, Minnesota, a rare stopover point for Steinheimer during an upper-Midwest trip in 1957.

**PLATE 113** Denver and Rio Grande Western trainman drains cylinder cocks, Durango, Colorado, 1961.

The ethereal glow from torchlight illuminates a railroad worker draining cylinder cocks on doubleheader Mikados outside the roundhouse in Durango, Colorado. The train is just in from Alamosa; the time is 1961, and the narrow gauge lines are on borrowed time.

**PLATE 114** Southern Pacific 2-10-2 #3625 and trainmen, Saugus, California, 1948.

In a posed photograph that combines time exposure and flash-fill, two enginemen linger beside their 2-10-2 locomotive on helper duty awaiting the arrival of the *Montalvo Local* coming off the Santa Paula Branch. Orange blossoms perfume the air as citrus production was the prime industry in the Santa Clarita Valley. The ungainliness of the 4 x 5 Speed Graphic Steinheimer then used often made set up shots mandatory. With the adaptation of smaller cameras later on, he eschewed this approach, relying more on a less intrusive, journalistic style when including people in his compositions.

**PLATE 115** Southern Pacific's *San Joaquin Daylight* at tunnel #3, Tehachapi, California, July 1952.

Looking more like a model railroad diorama than the real thing, Southern Pacific's *San Jaoquin Daylight* breaks into sunshine as it emerges from rebuilt tunnel #3 behind an Alco PA and Baldwin road-switcher one month after the devastating Tehachapi earthquake in Southern California.

**PLATE 116** Cab ride on the Oakland Terminal Railroad, Oakland, California, 1955.

In an early cab ride, Steinheimer stands on the gangway of an Oakland Terminal switch engine as it glides across a transfer bridge above the SP main line—a main line where cab-forward #4210 gets ready to head to Roseville with eighty cars on the drawbar. Steinheimer covered the varied aspects of railroading in the East Bay including Key System interurbans and Sacramento Northern boxcabs.

**PLATE 117** Denver and Rio Grande Western freight, Cumbres, New Mexico, 1961.

In a gambled exposure of 1/30 of a second, Steinheimer does a slow, handheld pan shot of a Rio Grande freight struggling upgrade toward the summit at Cumbres; sunset has fallen and a full moon is on the rise. While the narrow gauge would hold on until 1969, this view, made in 1961, suggests twilight will soon descend on the anachronistic steam operation in Colorado.

**PLATE 118** Milwaukee Road, moonlit portrait of Little Joe #E77, near Avery, Idaho, 1973.

The photographer rides the front gangway of a trailing SD40 to make a moonlit portrait of Little Joe #E77 leading train #264 out of Avery, Idaho, in 1973, just months before CMSt.P&P's complete electrification program ended. The pictures, taken with available light and twenty-four frames of Tri-X, represent a pinnacle of creativity and risk taking for Steinheimer; these images established new benchmarks for artistic possibilities in action railroad photography.

**PLATE 119** Chicago, Burlington and Quincy freight yard, East St. Louis, Illinois, November 1965.

An SD9 emblazoned with the CB&Q's ubiquitous "Way of the Zephyrs" script brings a cut of cars into a transfer yard in East St. Louis, a place not normally associated with Steinheimer. A truck, perhaps aiding in less-than-carload service, frames the foreground; a brakeman guides the locomotive through the switches.

**PLATE 120** Denver and Rio Grande Western's *California Zephyr* near Plainview, Colorado, 1964.

The Midwestern prairiescape stretches away to the east as the *California Zephyr* climbs the escarpment of the Rockies above Plainview, Colorado, on its 2,500-mile journey from Chicago to Salt Lake City, and finally the fabled Feather River Canyon in Northern California and Oakland. A Budd-built observation dome car brings up the markers on this streamliner inaugurated in 1949.

**PLATE 121** Southern Pacific, turntable and roundhouse with F3 #6164 and Krauss-Maffei #9010, Bakersfield, California, October 1964.

A cool Central Valley night envelops roundhouse whisker tracks as Southern Pacific F3 #6164 and Krauss-Maffei hydraulic locomotive #9010 await future road assignments. The turntable's considerable age shows: the tender's shanty and handbrake seem holdovers from an earlier industrial era. The experimental K-Ms, while favored on many Valley freights for their pulling power, would be banished from the roster just four years later in 1968, a 9-million-dollar gamble on alternative technology that failed to pay off. Bakersfield, too, was an important staging point for trains going north toward Portland or those heading east over the Tehachapi Mountains; the Central Valley's vast agricultural output also coalesced here from feeder lines strung across the San Joaquin. The photo was taken while Steinheimer was on a trip with friend Robert Knoll.

**PLATE 122** Hobo sleeping near Southern Pacific yards, Roseville, California, June 1964.

Reminiscent of a Dorothea Lange or John Vachon FSA image from the 1930s, Steinheimer records a member of the "hobo fraternity" resting on cardboard in the shade of an outbuilding at SP's Roseville Yard. Steinheimer, always inclusive and curious, never shied away from photographing the less-glamorous aspects of railroad culture.

**PLATE 123** Sierra Railroad railfan trip as seen from car, Northern California, late 1950s.

Merging automobility with railfanning, Steinheimer takes a photo of what is probably a Sierra Railroad fan trip from the backseat of his car in the late 1950s. He, along with other railfan photographers from his generation like Philip Hastings and Robert Hale, made thorough use of the car to extend their photographic creativity, not only to get them further afield but also as a means to make different types of photographs, most notably the "pacing" shot.

**PLATE 124** Southern Pacific's *City of San Francisco,* near Wells, Nevada, 1965.

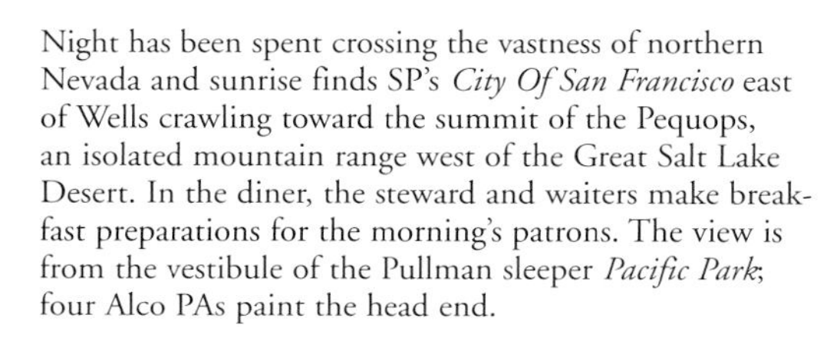

Night has been spent crossing the vastness of northern Nevada and sunrise finds SP's *City Of San Francisco* east of Wells crawling toward the summit of the Pequops, an isolated mountain range west of the Great Salt Lake Desert. In the diner, the steward and waiters make breakfast preparations for the morning's patrons. The view is from the vestibule of the Pullman sleeper *Pacific Park*; four Alco PAs paint the head end.

**PLATE 125** Southern Pacific freight westbound at Mount Hebron, California, 1960s.

The hurried passage of a Southern Pacific Oregon-bound freight, with a hood unit and two Fs, sends a fresh dusting of new-fallen snow skyward as it makes quick time through a whitened landscape near Mount Hebron, California, a lonely outpost on the Cascade line between Weed and Klamath Falls just south of Dorris, California.

**PLATE 126** Denver and Rio Grande Western engineman comes off duty, Salt Lake City, Utah, October 1952.

In one of Steinheimer's personal favorite photographs, a D&RGW engineman comes off duty at Roper Yard in Salt Lake City, the late-afternoon glare of a low-slung sun making for a squinty portrait. Steinheimer, early on, was attracted to including human interest in his imagery, a trait that set his work apart.

**PLATE 127** Southern Pacific's *City of San Francisco* in Cold Stream Canyon, near Andover, California, 1962.

Train #101, the *City of San Francisco*, rounds Stanford Curve in Cold Stream Canyon on SP's Donner Pass as it mounts the railroad's 1.9 ruling grade of the formidable 7,017-foot crossing of the Sierra Nevada. In 1952 this same train made national headlines when it was stranded near Yuba Gap by snowslides for three days.

**PLATE 128** Great Northern's *Empire Builder,* near Browning, Montana, May 1964.

Steinheimer employs a favorite framing device of rail photographers—the snowshed portal—as Great Northern FP7 #350 enters a wooden version near Browning, Montana, with the westbound *Empire Builder.* Completed in 1888, GN's main line was the most northern of the transcontinentals; at 5,228 feet it's also the lowest crossing of the Rocky Mountains by any railroad in the United States.

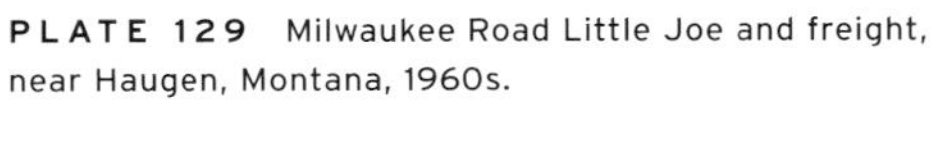

**PLATE 129** Milwaukee Road Little Joe and freight, near Haugen, Montana, 1960s.

In a display of contrast only a black-and-white printer could love, a heavy Bitterroot Range snowfall accumulates atop pine-studded forests, while a Milwaukee Road freight train led by Little Joe #E20 and two trailing EMD units rolls out of Haugan, Montana.

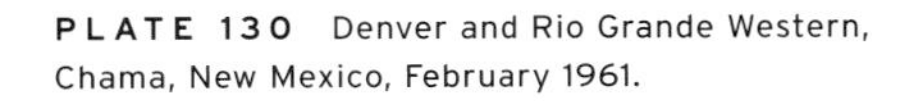

**PLATE 130** Denver and Rio Grande Western, Chama, New Mexico, February 1961.

In one of Steinheimer's more mysterious and surreal images, a whirl of steam obscures a engineman ready to board narrow gauge Mikado #494 in the yard at Chama, New Mexico.

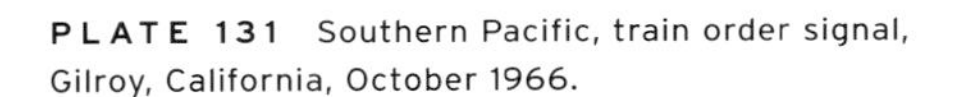

**PLATE 131** Southern Pacific, train order signal, Gilroy, California, October 1966.

The train order board at Gilroy, California, standing elegiac against high clouds, indicates to the approaching train that a "flimsie" awaits retrieval. The SP's Coast Line succumbed to radio control and DTC (Direct Traffic Control) implementation in the mid-1980s. Gilroy's station, eighty miles south of San Francisco, closed in 1971 and was renovated for CalTRANS commuter service in 1999. Steinheimer recorded the simple forms of this signal—a familiar railroad icon—in the fall of 1966.

**PLATE 132** Southern Pacific eastbound freight on Suisun Bay Bridge, Benicia, California, early 1950s.

Exploiting naturally occurring geometries of the railroad environment, Steinheimer frames the passage of an eastbound SP freight crossing the Suisun Bay Bridge in the early 1950s on its run toward Roseville, California. The mile-long bridge rendered ferry service (used since 1879 to transport trains between Port Costa and Benicia) unnecessary when it was completed in 1930. The *Solano* and *Contra Costa,* two of the largest car-transfer ferries ever built, made their final voyages across the Carquinez Straits in November of that year.

**PLATE 133** Union Pacific's *North Platte Valley Express,* Yoder, Wyoming, 1953.

An arresting tableau of rural railroading plays out before the camera as the UP's *North Platte Valley Express* with a 7,000-class 4-8-2 on the head end makes a momentary station stop at the tank town of Yoder, Wyoming, in 1953. Early on, Steinheimer frequently stepped back to include the mis-en-scène of the world of railroading.

**PLATE 134** Colorado and Southern northbound freight, above Horse Creek, Wyoming, 1956.

A northbound Colorado and Southern freight winds its way through the sweeping curves above Horse Creek just north of Cheyenne, Wyoming. Ahead is the Wind River Canyon and Laurel, Montana. The C&S was a subsidiary of the CB&Q.

**PLATE 135** Denver and Rio Grande Western meet between two freight trains, Price Canyon, Utah, 1951.

Two Rio Grande helpers send awe-inspiring displays of steam skyward as a coal drag meets an eastbound manifest piloted by F3 diesel intruders from EMD in the Price River Canyon between Helper and Soldier Summit, Utah. This shot was taken during the winter trip Steinheimer and Don Sims made over a five-day period in December 1951; they returned home with several iconographic masterpieces in their film holders.

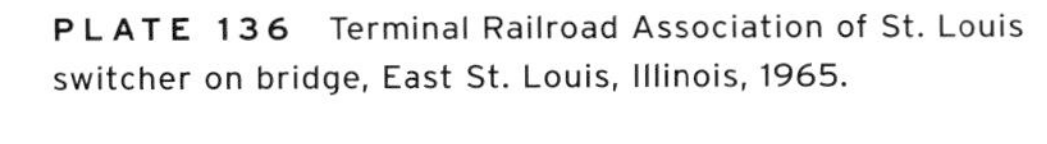

**PLATE 136** Terminal Railroad Association of St. Louis switcher on bridge, East St. Louis, Illinois, 1965.

In East St. Louis, Illinois, a location not normally associated with Steinheimer, the photographer captures a Terminal Railroad Association of St. Louis switch engine crossing Gulf, Mobile and Ohio trackage during a 1965 business trip for Fairchild Semiconductor. The TRRA, at one time jointly owned by all railroads within St. Louis, switched Union Station in its heyday and owned the famed Eads Bridge and Eads Tunnel found beneath the city's streets.

**PLATE 137** Southern Pacific cab-forward #4363 and section man on speeder, Saugus, California, early 1950s.

Train #60, the *West Coast,* led by cab-forward #4363—a locomotive unique to the SP—dwarfs a nearby section-man patiently awaiting its passage while seated on his speeder. The location is Saugus, California, on the SP's Valley line, and the train is destined for an 8:50 AM arrival at Los Angeles Union Passenger Terminal. The time is the early 1950s, and steam still has a dwindling hold on Southern California.

**PLATE 138** Southern Pacific freight with U25B #6765, near Dos Cabezas, California, 1967.

An eastbound freight with GE-built U25B #6765 emerges from SD&AE's tunnel 21, amid the boulder-strewn landscape of the rugged Jacumba Mountains in Southern California. Construction of the original San Diego & Arizona, instigated by sugar magnate John D. Spreckels (with a three-million-dollar loan from E. H. Harriman and the Southern Pacific) in order to tap the agricultural crops of the Imperial Valley and give San Diego a direct link to transcontinental main lines, began in 1907, and was completed to El Centro in 1919. After years of financial struggle, however, the Spreckel's family sold the line to the SP in 1933, which renamed the railroad the San Diego and Arizona Eastern. Collapsing tunnels and devastation wrought by Hurricane Kathleen caused the SP to abandoned the line in 1975.

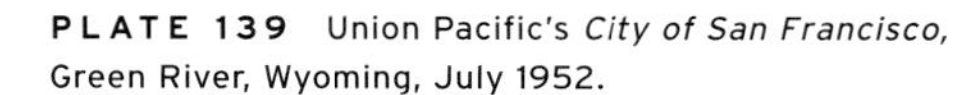

**PLATE 139** Union Pacific's *City of San Francisco,* Green River, Wyoming, July 1952.

A tardy *City of San Francisco* stops for passengers in Green River, Wyoming. An E9 leads the transcontinental streamliner inaugurated in 1938 and jointly operated by the C&NW, UP, and SP. The high plateau country of southwestern Wyoming surrounds the train. Steinheimer predecessors W. H. Jackson and Timothy O'Sullivan photographed many of the geologic formations in the area, including the well-known Castle Rock, an early landmark for emigrants on the Overland, Oregon, and Mormon trails as they headed west.

**PLATE 140** Southern Pacific's *Lone Pine Local,* Cantil, California, 1971.

A lone SD9E with 4 cars approaches Cantil tank on its daily run up the Jawbone Branch to Lone Pine, California. This 139-mile line left the SP main at Mojave, running near barren, beautiful geologic features like Red Rock Canyon and through the Owens Valley. At one time, it formed a vital link at Owenyo to the fabled Carson and Colorado, a narrow gauge line strung through the desert to mineral reserves of the Cerro Gordo and other mines around the Inyo Mountains. The last *Local* left Lone Pine in 1982 with the rails pulled up behind it soon thereafter, choking off its northern section. Today a remnant of the branch from Mojave to Searles survives with potash trains off the Trona Railway.

**PLATE 141** Southern Pacific commute train with 4-8-4 #4405, South San Francisco, California, mid-1950s.

Southern Pacific GS #4405 and a dozen heavyweight suburban coaches slip on wet rail as commute train 123 gathers momentum departing South San Francisco on a foggy morning in the early 1950s. Ahead is the Bayshore Cutoff, a 1904 Coast Route track realignment instituted during the Harriman era. The ten-mile project cost over nine million dollars, saw five tunnels bored, and was completed in December 1907. Commuter ridership peaked at 16,000 round-trips per day in the mid-1950s.

**PLATE 142** Southern Pacific, laborer washing cab windows on *City of San Francisco,* Oakland, California, 1966.

A favorite subject of Steinheimer's in the 1960s was train #101, the *City of San Francisco,* shown here getting a quick "makeover" between runs at Oakland's 16th Street Station. A laborer washes the Alco PA's windows before its 2:50 PM departure for Chicago. A well-known Steinheimer article on PAs and the *City* ran in *Trains* in November 1967.

**PLATE 143** Southern Pacific, Taylor Yard Diesel Facility, Los Angeles, California, 1949.

Steinheimer positions his 4 x 5 Speed Graphic low, practically in the servicing pit, to accent the monumentality of the then-new E9, being inspected by a workman at Southern Pacific's Taylor Yard diesel shops. The E9s were routinely used on the Golden State and Sunset Route passenger trains out of Los Angeles and featured a brilliant red and orange paint scheme.

**PLATE 144** Atchison, Topeka and Santa Fe depot interior and operator's desk, Palmer Lake, Colorado, April 1964.

Still life at Palmer Lake, Colorado. Steinheimer makes a interior portrait of the Santa Fe depot that captures the quiet visual wealth of railroad's trappings and traditions: the scissor phone, turnover calendar, typewriter, telegraph sounding key, standard clock, and semaphore levers. True to the layered approach to his work, he made this image moments after documenting the passage of the Missouri Pacific's *Colorado Eagle* (see plate 76) during a whirlwind three-week trip around the intermountain West in 1964. For Steinheimer, every aspect of the railroad environment received democratic treatment on his contact sheets; everything warranted film time.

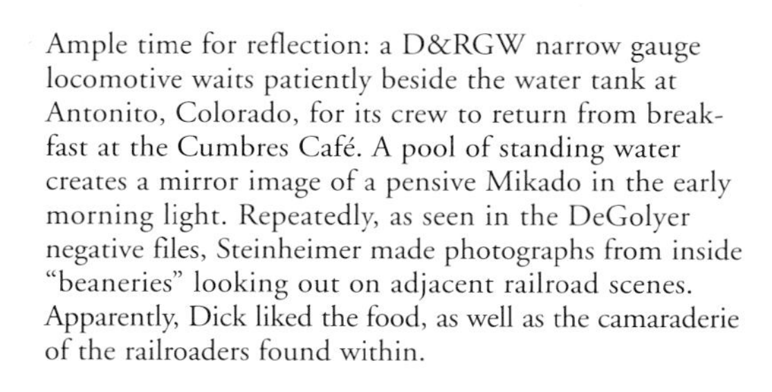

**PLATE 145** Denver and Rio Grande Western, Mikado and reflection, Antonito, Colorado, February 1961.

Ample time for reflection: a D&RGW narrow gauge locomotive waits patiently beside the water tank at Antonito, Colorado, for its crew to return from breakfast at the Cumbres Café. A pool of standing water creates a mirror image of a pensive Mikado in the early morning light. Repeatedly, as seen in the DeGolyer negative files, Steinheimer made photographs from inside "beaneries" looking out on adjacent railroad scenes. Apparently, Dick liked the food, as well as the camaraderie of the railroaders found within.

**PLATE 146** Denver and Rio Grande Western, Mikados and freight, near Falfa, Colorado, February 1961.

2-8-2 Rio Grande Mikados #483 and #494 roll past a graveyard of Chevrolets, Buicks, and Oldsmobiles on their climb toward Falfa, Colorado, with almost eighty empties on the drawbar.

**PLATE 147** Denver and Rio Grande Western 2-10-2 #1403 with coal drag, Thistle, Utah, 1951.

At Spanish Fork Canyon just outside Thistle, Utah, in 1951, Steinheimer makes one of his early steam-era masterpieces. A D&RGW drag freight of empty coal cars and gondolas heads for Soldier Summit. This image, a companion piece to the more frequently reproduced one in plate 7, was the second sheet of 4 x 5 film Steinheimer exposed as this train passed his camera position. The final image was made as the caboose went by.

**PLATE 148** Atchison, Topeka and Santa Fe *San Diegan* prior to departure, San Diego, California, February 1952.

Budd/ACF car #77 stands waiting in San Diego for a 7:15 PM departure to Los Angeles. A car inspector, looking over the train, stops to chat and discuss last-minute details with a flagman prior to "train time." The unique "pendulum" cars, regulars on AT&SF's *San Diegans,* were harder to repair but rode better than conventional equipment and banked around curves, maintaining equilibrium for the passengers inside. This was just one experiment by a railroad dedicated to passenger service and whose name was synonymous with first-class train travel.

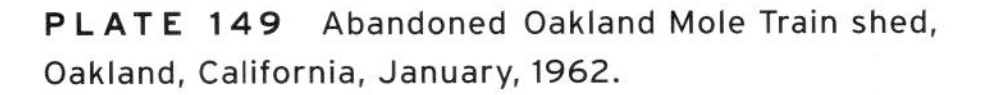

**PLATE 149** Abandoned Oakland Mole Train shed, Oakland, California, January, 1962.

Wearing its eighty-year history wearily, the entrance to Oakland Mole stands in a state of decay. Cultural geographer J. B. Jackson once said that as a people we only truly understand the value of a building after it's in ruin or been demolished. The Mole had outlived its usefulness by 1958, and was supplanted by automobility, two bridges, and a bus service that drove passengers across the Bay. As Steinheimer recorded a lasting visual memory of a place dear to him, an elderly woman pauses to wonder over its remnants. The trainshed succumbed to the wrecking ball in 1966, unsaved by the nascent preservation movement. Here, it's savored one last time by a photographer who'd known its spaces intimately since the early 1950s.

**PLATE 150** Southern Pacific, disabled F7 units on turntable, Norden, California, January 1967.

A lone mote of daylight pierces the darkness of Norden's labyrinthian snowsheds, as a disabled Southern Pacific F7A and B-unit ride the covered turntable—itself a rarity in railroad architecture—before returning, dead in train, to Roseville, California, for repairs.

**PLATE 151** Southern Pacific Burbank Junction Tower, Burbank, California, 1950

Burbank Junction will soon rumble to the passage of the Bakersfield-bound "Valley" train #807, with F7 #6146 in the lead. The camera location, with its "pulled back" vantage point, reveals a wealth of information about the San Fernando Valley railroad corridor: interlocking plant, late-model cars, standard SP tower, train order stand with bam rods and flimsies attached, and a busy thoroughfare.

**PLATE 152** Southern Pacific freight with SD45 #8845, Lodi, California, 1971.

As in the Credence Clearwater Revival song lyric "Stuck in Lodi again," the SP's San Joaquin Valley Route south of Sacramento seems mired in a prolonged dense tule fog. SD45 #8845 South, crawls through town on this February day in 1971. The Lodi depot, constructed in 1907, was not the usual SP type #22 but instead featured a colonnade architectural style—one of about thirty constructed from redwood by the SP between 1901 and 1913.

**PLATE 153** Southern Pacific F-M Trainmaster #4813 and commute train, Palo Alto, California, 1964.

For over a year, Steinheimer repeatedly climbed the rooftop of his suburban home in Palo Alto to make pan shots of Southern Pacific's passing late-afternoon commute "parade." Initial attempts were often blurred or chaotically composed; this, however, was unfettered and experimental self-expression. While he eventually succeeded (if "success" is defined as a sharp locomotive and blurred background), some of the outtakes, like this image of Trainmaster #4813, with jagged shadows of backyard and fence, are intrinsically more interesting.

## A Word About the Photographs

While it is normally customary in a fine-art photography book to run uncropped photographs as originally recorded by the artist, in this case, after viewing many original Steinheimer negatives at the DeGolyer Library and seeing subsequent prints of those same negatives, I came to realize that Dick actually shot "fat"—meaning he intentionally left room for cropping, especially during his 4 x 5 phase—perhaps as a result of his newspaper background. Also, while combing through his existing print files in Sacramento it wasn't unusual to come across "variants": different croppings of the same image, sometimes horizontal, sometimes vertical. Furthermore, he was used to sending images to trade journals, book projects, and magazines such as *Trains*, where photo editors and graphic designers routinely cropped images in a manner best suited to their needs. With this book, while I've tried to maintain a "full-frame" notion in reproducing the majority of images, I'm not immune to aesthetic prejudices as an editor, nor can I say my own sensibilities as a photographer haven't influenced cropping decisions a bit. Therefore, if any license has been taken with the photographs, it's been done solely to enhance the inherent graphic quality of Steinheimer's work. —JB

## Acknowledgments for Richard Steinheimer

My career in photography goes back to 1941, when my mother, Frances Julian, gave me the money to buy a three-dollar Kodak Baby Brownie camera. I'm indebted to her for giving a shy kid the needed push to come out of my shell; to cousin David Steinheimer, who first taught me how to develop film: I recall the eager anticipation as we watched railroad images appear in the developing trays. During the ensuing years my friend Don Sims, a seasoned railroad photographer, taught me to throw caution to the wind and not be afraid to take chances. I'm indebted to my first wife, Nona Steinheimer, who held down the fort, taking care of our children—Marilyn, Sally, and Alan—while I chased my dreams. I also want to thank my present wife, Shirley Burman, who persistently encouraged me to bring my black-and-white photography out of storage, after I'd decided to leave it behind to pursue the medium of color. There are so many friends to acknowledge that it would take pages to list them all. More important though, this book would not have come to realization without the support and love of Wendy and Jeff Brouws. To them, Shirley and I are forever indebted.

## Acknowledgments for Jeff Brouws

*A Passion for Trains: The Railroad Photography of Richard Steinheimer* has been a project under construction for a long time. Initial talks with Dick and Shirley back in 1991 about a fine-art book showcasing Dick's extensive archive of black-and-white photographs became a decade-long conversation. Thankfully, the talk eventually turned to reality.

First and foremost, I want to thank Dick Steinheimer for welcoming me into his home almost thirty years ago. That initial two-hour visit in Palo Alto, California, was life altering and fostered notions that I, too, might go on to a career in photography. His selfless encouragement to legions of younger photographers over the last fifty years has rightfully made him a hero to many. But to be inspired by a photographer's work is one thing; to have access to him as a human being—to enjoy his friendship and camaraderie—is another, a rare occurrence. He's a gentle, humble, loving person, and that too has been lesson bearing and life affirming. Thank you, Dick, for this openness.

Second, I would also like to acknowledge, in the fullest definition of that word, Dick's wife, Shirley Burman, for all the energy, enthusiasm, and sheer determination she lent to this project. Without her unstinting good nature and willingness to help pull together the disparate elements that comprise the skeletal backbone of any book project, the publication you have in your hands might not have come into existence. She answered every phone call and e-mail request, aided in organizing the negatives we borrowed from the DeGolyer Library (and oversaw the necessary interface required there with its staff), orchestrated getting the prints done from those negatives, and found obscure bits of ephemera from Dick's archives that helped in rounding out the essay material. She was tireless, always available to help, while selflessly putting her own photography and writing projects on hold. Shirley did all this while simultaneously beginning to catalogue and reorganize the DeGolyer Steinheimer Collection on behalf of that institution.

I also need to send out a grand thanks to John Gruber, another angel on our shoulders for this project. I received countless envelopes containing xeroxs that helped me gather necessary material for my text on Steinheimer and his relationship to other cameramen, past and present, who also courted trains. John has spent years unearthing little-known facts about the evolution of railroad photography in America, and I've found his own research and work in this area—and readiness to share it—an ongoing source of inspiration.

Thanks also go to Ted Benson—photojournalist, artist, writer—for his excellent work and previous scholarship surrounding Steinheimer's life. The many articles about Dick that Ted wrote for *Locomotive and Railway Preservation* and *Vintage Rails* magazines not only provided me with detailed background information, but this research, covering certain aspects of Steinheimer's career, freed me to take a slightly different tack and look at other facets of Steinheimer's lifelong engagement with trains. Ted's groundbreaking work made my job much easier, and for this I am indebted.

A special thanks as well to Kalmbach Publishing Company and *Trains* editors Mark Hemphill, Kevin Keefe, and Rob McGonical for their decision to grant the *Trains* 2002 Preservation Award to the DeGolyer Library (at Southern Methodist University in Dallas) to help organize the Steinheimer archives; a truly magnanimous gesture.

I'd also like to extend a heartfelt thanks to our editor at W. W. Norton Jim Mairs. Jim's former association with another great railroad photographer, Don Ball, made him the perfect liaison to guide this project through its various stages—his was a gentle hand on the "throttle." I also think his love of trains, and his early enthusiasm for Steinheimer's images, brought something additional to these pages; it's been a pleasurable collaboration. Our other friend at W. W. Norton, Bill Rusin, and his passionate love of photography in general, proved instrumental in bringing Steinheimer's work to a larger audience too. Thanks also to our nimble copy editor, Brook Wilensky-Lanford.

Special thanks to the staff at the DeGolyer Library at Southern Methodist University in Dallas who made access to Steinheimer's files possible: Director Russell Martin and administrative assistant Betty Friedrich, Francisco Garcia, Kathy Rome, Clint Scott, and Kammie Schuman. Our week in Dallas in 2002 went smoothly due to your generosity and good cheer.

I also want to thank my wife, Wendy Burton Brouws, for her contributions to the project. She selflessly spent countless hours at the DeGolyer Library by my side scanning negatives and typing caption material. The excellent sequencing of images herein is also her creative handiwork. She acted as book agent, and got the necessary materials into the right hands at W. W. Norton, the genesis point for any set of photographs or manuscripts hoping to see publication. The book truly wouldn't have happened without her.

There were also countless other individuals and organizations that gave interviews, lent important research help, granted permission to reprint text or photographs, or otherwise aided our efforts; they are gratefully acknowledged here: Wallace Abbey, the Amon Carter Museum, Burlington Northern Santa Fe Railway, B. B. Burton, S. R. Bush, George Cook, Ed Delvers, Wayne Depperman, Terry Falke, Chuck Fox, Cornelius Hauck, Kentaro Hirai, Anne Clegg Holloway, Ron Hill, Japan Times, Ltd, Joel Jensen, Bruce Kelly, Greg McDonnell, MIT Press, Mel Patrick, John Pickett, Ken Rehor, Gil Reid, Jim Shaughnessy, Jane Tompkins, *Trains* magazine, Stephen White and the Stephen White Photography Collection II, Persia Wooley, and the Whitney Museum of American Art.